SIMPLIFIED SITE
ENGINEERING

Other titles in the Parker–Ambrose Series of Simplified Design Guides

Harry Parker and James Ambrose
Simplified Design of Concrete Structures, 7th Edition

Harry Parker, John W. MacGuire and James Ambrose
Simplified Site Engineering, 2nd Edition

James Ambrose
Simplified Design of Building Foundations, 2nd Edition

James Ambrose and Dimitry Vergun
Simplified Design for Wind and Earthquakes Forces, 3rd Edition

Harry Parker and James Ambrose
Simplified Design of Steel Structures, 6th Edition

James Ambrose and Peter D. Brandow
Simplified Site Design

Harry Parker and James Ambrose
Simplified Mechanics and Strength of Materials, 5th Edition

Marc Schiler
Simplified Design of Building Lighting

James Patterson
Simplified Design for Building Fire Safety

James Ambrose
Simplified Engineering for Architects and Builders, 8th Edition

William Bobenhausen
Simplified Design of HVAC Systems

James Ambrose and Jeffrey E. Ollswang
Simplified Design for Building Sound Control

James Ambrose
Simplified Design of Building Structures, 3rd Edition

SIMPLIFIED SITE ENGINEERING

Second Edition

THE LATE HARRY PARKER, M.S.

Formerly Professor of Architectural Construction
University of Pennsylvania

and

THE LATE JOHN W. MACGUIRE, B. ARCH.

Formerly Professor of Architectural Engineering
University of Pennsylvania

prepared by

JAMES AMBROSE, M.S.

Formerly Professor of Architecture
University of Southern California

JOHN WILEY & SONS, INC.

New York • Chichester • Weinheim • Brisbane • Singapore • Toronto

This text is printed on acid-free paper.

Copyright © 1991 by John Wiley & Sons, Inc.

All rights reserved. Published simultaneously in Canada.

Library of Congress Cataloging-in-Publication Data

ISBN 0471-52809-9 (cloth)
ISBN 0471-17987-6 (paper)

SKY10021278_091720

CONTENTS

PREFACE

This new edition of Parker and MacGuire's 1954 book preserves and updates what has proven to be a highly useful reference for the introduction of basic materials relating to the engineering design of sites. The principal attribute of this book has been its conciseness and its easy use by persons with minimum education in mathematics and little background in engineering.

This redevelopment of the book seeks to preserve the essential character of the first edition, as described by the original authors in the Preface to the First Edition, which follows. Most of the work in the first edition has been retained, except for the extensive use of logarithms, now made obsolete by both the computer and the simple pocket calculator.

Where appropriate, materials have been brought up to date. Surveying work is now done routinely with equipment and procedures immensely more sophisticated than those available when this book was originally conceived, almost 40 years ago. Surveying

work is now done almost exclusively by professional, registered, Land Surveyors, using the latest equipment and procedures. The work itself remains essentially unchanged, but the processes for achieving it are substantially changed.

However, this book was never intended to be a thorough reference for training of professional surveyors; hence the first word of the title. Its basic purpose was—and is in this edition—the introduction of the most fundamental problems of site engineering to persons with little training in the subject. For this purpose, the treatment is reduced to the simplest and most fundamental issues and tasks. Thus, it is considered to be more important for the reader to understand *what* is being considered and *why* it is important in the work discussed, than how it can be achieved by the latest means.

Although professional surveying work is now done with more elaborate equipment, the simple builder's level and transit are still capable of being used to perform the most basic tasks of surveying. In fact, these instruments still exist, can still be purchased new, and are relatively easy to learn with and use for some minor surveying tasks. The examples in this edition therefore retain the illustrations of work with these simple instruments. Once the concepts of what is being done are in hand, using more modern equipment to do the job would be a logical step for those wishing to seriously pursue the work of professional surveying.

While this book may be an easy first step for those persons who intend to become professional surveyors, its real purpose is to introduce the topic for architects, civil engineers, landscape architects, and builders; none of whom are likely to actually *do* much surveying work in their developing careers. The original title of the book, in fact, was *Simplified Site Engineering for Architects and Builders*. This was the seventh volume of such a nature produced by Professor Parker; the first six being on the subject of structures (not really intended for persons pursuing professional careers in structural engineering).

Some new materials have been added in this edition, pursuing the issues of site development into more concerns related to landscape development and building construction on sites. One of the purposes in developing this edition has been to make the book one of a three-volume set on general site-related design concerns. The other two volumes are:

Simplified Design of Building Foundations, 2nd ed., James Ambrose, Wiley, New York, 1988.

Simplified Site Design, James Ambrose and Peter Brandow, Wiley, New York, 1991.

Each of these books treats issues and design problems related to site development from a slightly different point of view. Still, the subject has many facets, and each somewhat overlaps the other; a symptom of the professional world, where the work of architects, civil engineers, structural engineers, planners, and landscape architects tends to overlap. The basic focus of the books is essentially different, however, giving some basis for concentration of the book scope and treatment. This book basically develops topics of prime concern to Civil Engineers and Land Surveyors. The foundations book develops topics of prime concern to Architects, Structural Engineers, and Geotechnic Engineers. The site design book develops topics of prime concern to Planners, Landscape Architects, and Architects.

I must acknowledge the fine work of the late Harry Parker and John MacGuire in their concise and precise development of the first edition; most of which is preserved here intact. I am also grateful to the editors and production staff at John Wiley & Sons for their support and their usual competent and reliable work. Finally, I am grateful to my wife, Peggy, for her assistance with typing and proofreading of the work and with the myriad tasks of finishing the book, and to my son, Jeff, for his assistance with the graphic work.

JAMES AMBROSE

Westlake Village, California
August 1991

PREFACE TO THE FIRST EDITION

In the preparation of a set of architectural drawings for a building project, numerous problems on many subjects must be solved. Many of these are routine procedure and can be solved by any architect. There is, however, a certain type of problem that requires the attention of someone who has specific knowledge of problems that arise in the analysis of building sites and the preparation of the site plan. Many architectural offices lack men who are qualified for this work, and this book has been written to explain and solve problems of this type. Those who have the knowledge afforded by a course in surveying are equipped to handle many of the problems that are encountered.

This, however, is not just another book on surveying, and no attempt has been made to describe all the many methods of procedure and computations. Simple methods for using surveying instruments are explained; they may be readily applied. Although most architects are seldom required to perform actual surveying, they frequently need the ability to perform the office computations that result from surveying data. This book explains in detail the solution of problems of this nature, and numerous examples are presented. The book will serve also as a review and refresher on subjects that may have been forgotten.

The items contained are both numerous and varied, and the following list enumerates some the problems that are discussed, illustrated, and solved:

The interpretation of deed descriptions from which site plans are plotted.

The dimensioning of buildings and sites when the angles are other than right angles.

The operation of surveying instruments and the making of surveys and site plans.

The computation of areas of irregular plots.

Dimensioning and laying out of circular curves for driveways and buildings involving acs of circles.

Vertical curves.

The analysis of contour lines and their manipulation in the solution of grading problems.

The computation of excavation volumes in ground having an uneven surface.

The computation of volumes of cut and fill as indicated by contour lines.

The use of the planimeter.

Maximum and minimum grades for driveways, sidewalks, play areas, etc.

The computation of drainage pipe sizes.

Staking out buildings and driveways.

Items to be considered in the selection of a site.

A check list for site plans.

In the solution of mathematical expressions one can, if he so desires, multiply and divide by the methods used in everyday arithmetic. But, in the dimensioning of plans and sites, the use of logarithms affords both a high degree of accuracy and a great saving of time. To familiarize the user of this book with this valuable tool, the basic principles of logarithms are explained and a five-place log table of numbers is provided. Trigonometric functions are included, with explanations of their uses when required in the dimensioning of drawings. One need have no knowledge of advanced mathematics to understand the computations included in this book; arithmetic and high-school algebra afford adequate preparation. A unique feature of this book is that problems requiring computations are accompanied with logarithmic computations shown in detail.

Following the format of the other books in the "Simplified Series," this book contains concise explanations of procedures illustrated by the solution of practical problems. Included also are problems to be solved by the student. Because of the manner of presentation and the arrangement of material, this book is appropriate for use in classrooms as well as for home study.

Thanks are extended to the Keuffel and Esser Company for permission to adapt from their catalogues the cuts of instruments shown in Figures 5.1, 5.2, and 11.5.

The authors are well aware that many of the problems relating to site engineering demand the services of a qualified engineer. There are, however, many problems of this nature that arise constantly that may be readily solved by the architect or builder. The purpose of this book is to provide the knowledge for such problems.

HARRY PARKER
JOHN W. MACGUIRE

Philadelphia, Pennsylvania
July 1954

1

INTRODUCTION

1.1 SITE DEVELOPMENT

This book treats the engineering investigations and design work related to the development of sites for specific purposes. The primary purpose treated here is that of sites intended for construction of buildings. In special situations, sites may be left in essentially natural conditions, except for the intrusion of the building construction. Typically, however, sites must be developed (or redeveloped) to facilitate the purposes of the users of the buildings.

Redevelopment of sites is basically a two-stage process. First, the existing site conditions must be thoroughly investigated and the information recorded for use by designers and builders. Then the necessary modifications of existing conditions must be planned. Modifications are done to facilitate some intended use of the site, and when this involves the construction of buildings, the site work will be closely related to the general development of the building plans.

1.2 CONCERNS FOR SITE DEVELOPMENT

There are typically many people involved in any site development. These include the owner or purchaser of the site, the various designers who work with the site, the contractors who do site and building construction work, and the local authorities and agencies who control the type of work planned for the site.

Work for site development typically overlaps the interests of many designers and consultants (see Figure 1.1). Site design activities are not always clearly divided between the major design professionals, and the work often overlaps their spheres of action. For small projects, the primary design profes-

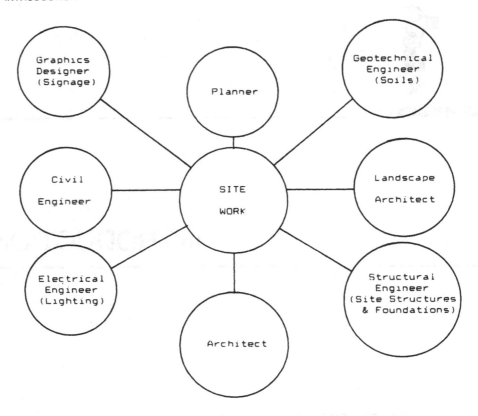

FIGURE 1.1. Potential contributors to site design.

sional (usually an architect or civil engineer) may do all or almost all of the site design work. On large projects, however, the work is usually divided between many specialists, requiring considerable coordination and management of the design work.

1.3 DIVISION OF SITE DESIGN WORK

Activities related to site development typically include the following.

Surveys and Investigations

This includes a thorough documentation of the existing surface conditions and site features, a subsurface investigation of site materials and geology, and any indicated special studies for seismology, water conditions, slope stability, and so on.

Site Engineering

This includes the general recontouring of the site surface, design for control of site drainage, planning for site construction (pavements, retaining walls, etc.), general planning for utilities and services (sewers, water, and other underground elements), and development of the site edges to relate to existing streets and properties.

Landscape Design

General usage and decorative development of the site, including use of plantings and other site materials.

Foundation Design

The general design for the below-grade portions of the buildings, and for the building supporting foundations.

Design for Site Construction Work

The necessary planning for excavation, temporary shoring, dewatering, and other work related to achieving the site and below-grade construction work.

1.4 SCOPE OF THIS BOOK

This book essentially deals with the scope of work generally viewed as that done by civil engineers. This consists primarily of the work described under the category of "Site Engineering," in the preceding section. However, the subject treatment here is also extended into the area of investigative work, an understanding of this is essential to any design work involving the site.

Surveying, both for preliminary site investigation and for construction work, is now typically done by specialists (*land surveyors*): whereas in times past, it was mostly done by employees of civil engineering or contracting firms. However, a brief study of surveying methods and activities is a valuable learning experience for anyone involved in site design work. A considerable portion of the work in this book is devoted to the work of land surveying and the various applications of the data derived from it.

1.5 RELATION OF OTHER BOOKS

As has been mentioned in preceding discussions, the activities related to site design overlap many concerns and the work of various designers and contractors. This book has been developed in this edition to relate to two other books that deal with related topics.

These are:

Simplified Design of Building Foundations, 2nd ed., James Ambrose, Wiley, 1988.

Simplified Site Design, James Ambrose and Peter Brandow, Wiley, New York, 1991.

These three books have considerable overlap of topic concerns, although each has a particular focus. Special topics of particular concern to one of the volumes are given less in-depth treatment in the other volumes, which reference the volume with the more developed material. Consequently, citations of the other two books are occasionally made in this book to avoid the need for extensive treatment here.

Many other references are also cited throughout the text. A general list of the source materials for work here is given in the References section at the end of the book.

This book is intended as an introduction to the general work of site engineering. While its coverage is broad in that regard, its treatment is quite simple and basic. This can be a good first experience for those interested in pursuing professional careers in civil engineering or land surveying, but the work here is more specifically developed for those with other basic interests who have a need to acquire some understanding of site engineering.

In the interest of presenting work that can be pursued by those with limited mathematical backgrounds, the book presents some basic development of the simple work in trigonometry and geometry that is applicable to the illustrated computations shown in the book. This treatment of mathematics assumes a background of only simple high-school algebra and geometry. Readers with more mathematics preparation may find this a useful review, but can probably move on more quickly to the applied work in later chapters.

Readers with little mathematics preparation, however, are strongly urged to spend the time to study the early chapters in order to develop confidence with the relationships and procedures that are used in the practical problem applications in later work. Although slowly and carefully developed, the amount of mathematical material is really quite brief, and is specifically limited to the basic preparation needed for actual work in the problems in the book.

1.6 COMPUTATIONS

Computational work here is done by "hand" methods, whereas work in professional offices is now routinely done with computer-aided methods. This book is not intended as an office training manual, but as an opportunity for involvement in basic problems. The more shortcuts (computers, tables, etc.) between the reader and the computational work, the more obscure the processes become. Laborious hand calculations are deliberately pursued here as learning involvements. All the actual work presented here can be done on a simple pocket calculator of the "scientific" variety, one capable of simple trigonometric and logarithmic operations.

1.7 UNITS

Measurements of length, area, and volume in this book are calculated in feet and inches and various correlated units such as yards, miles, and acres. Conversions between these units and the metric-based units of the SI (Systéme International) system must frequently be performed in doing professional design work. Persons seeking to obtain work in site engineering should expect to encounter these problems and to become familiar with the necessary procedures for conversions.

Notation of quantities for length, area, and volume may be done in different ways. In this book, two methods are used to indicate lengths. The first uses feet and decimal parts of a foot: thus, four and one-half feet is noted as 4.5 ft. This form is generally used for computational work and on most engineering drawings.

The second method of linear measurement uses feet and inches and fractional parts of inches: thus, four and one-half feet is noted as 4 ft and 6 inches; frequently recorded with shorthand notation as 4′ 6″. This form of notation is frequently used on architectural and construction drawings. Conversion between these methods is discussed in Section 4.8.

Angular measurement in this book is done with the angle-degree-minute system for recording information on drawings. However, for computations, it is usually necessary to convert to a decimal fraction system, indicating fractions of one degree as a decimal; thus, an angle of 12 and one-third degrees is expressed as 12.33 degrees. Older surveying equipment, maps, and data references tend to use the degree-minute-second system, while newer work is mostly done with the decimal fractions, which work much more directly with computational procedures. A familiarity with both systems is necessary and, in most cases, conversions are quite simple.

SI Units

At the time of preparation of this edition, the building industry in the United States is still in a state of confused transition from the use of English units (feet, pounds, etc.) to the new metric-based system referred to as the SI units. Although a complete phase-over to the SI system seems inevitable, at the time of this writing most suppliers of construction materials and products in the United States are still largely resisting it. (The old system is now more appropriately called the U.S. system because England no longer uses it.)

For readers who need to make conversions between the two systems, there are three tables provided here. Table 1.1 lists the standard units of measurement in the U.S. system with the abbreviations generally used in this book and a description of the use in engineering work. In similar form, Table 1.2 gives the

TABLE 1.1 UNITS OF MEASUREMENT: U.S. SYSTEM

Name of Unit	Abbreviation	Use
Length		
Foot	ft	Large dimensions, building plans, beam spans
Inch	in.	Small dimensions, size of member cross sections
Area		
Square feet	ft^2	Large areas
Square feet	in.^{2}	Small areas, properties of cross sections
Volume		
Cubic feet	ft^3	Large volumes, quantities of materials
Cubic inches	in^3.	Small volumes
Force, Mass		
Pound	lb	Specific, weight, force, load
Kip	k	1000 lb
Pounds per foot	lb/ft	Linear load (as on a beam)
Kips per foot	k/ft	Linear load (as on a beam)
Pounds per square foot	lb/ft^2, psf	Distributed load on a surface
Kips per square foot	k/ft^2, ksf	Distributed load on a surface
Pounds per cubic foot	lb/ft^3, pcf	Relative density, weight
Moment		
Foot-pounds	ft-lb	Rotational or bending moment
Inch-pounds	in.lb	Rotational or bending moment
Kip-feet	k-ft.	Rotational or bending moment
Kip-inches	k-in.	Rotational or bending moment
Stress		
Pounds per square foot	lb/ft^2, psf	Soil pressure
Pounds per square inch	lb/in.^{2}, psi	Stresses in structures
Kips per square foot	k/ft^2, ksf	Soil pressure
Kips per square inch	k/in.^{2}, ksi	Stresses in structures
Temperature		
Degrees Fahrenheit	°F	Temperature

TABLE 1.2 UNITS OF MEASUREMENT: SI SYSTEM

Name of Unit	Abbreviation	Use
Length		
Meter	m	Large dimensions, building plans, beam spans
Millimeter	mm	Small dimensions, size of member cross sections
Area		
Square meters	m^2	Large areas
Square millimeters	mm^2	Small areas, properties of cross sections

TABLE 1.2 *(Continued)*

Name of Unit	Abbreviation	Use
Volume		
Cubic meters	m^3	Large volumes
Cubic millimeters	mm^3	Small volumes
Mass		
Kilogram	kg	Mass of materials (equivalent to weight in U.S. system)
Kilograms per cubic meter	kg/m^3	Density
Force (Load on Structures)		
Newton	N	Force or load
Kilonewton	kN	1000 newtons
Stress		
Pascal	Pa	Stress or pressure (1 pascal = 1 N/m^2)
Kilopascal	kPa	1000 pascal
Megapascal	MPa	1,000,000 pascal
Gigapascal	GPa	1,000,000,000 pascal
Temperature		
Degrees Celsius	°C	Temperature

TABLE 1.3 FACTORS FOR CONVERSION OF UNITS

To Convert from U.S. Units to SI Units, Multiply by:	U.S. Unit	SI Unit	To Convert from SI Units to U.S. Units, Multiply by:
25.4	in.	mm	0.03937
0.3048	ft	m	3.281
645.2	in.^{2}	mm^2	1.550×10^{-3}
16.39×10^3	in.^{3}	mm^3	61.02×10^{-6}
416.2×10^3	in.^{4}	mm^4	2.403×10^{-6}
0.09290	ft^2	m^2	10.76
0.02832	ft^3	m^3	35.31
0.4536	lb (mass)	kg	2.205
4.448	lb (force)	N	0.2248
4.448	kip (force)	kN	0.2248
1.356	ft-lb (moment)	N-m	0.7376
1.356	kip-ft (moment)	kN-m	0.7376
1.488	lb/ft (mass)	kg/m	0.6720
14.59	lb/ft (load)	N/m	0.06853
14.59	kips/ft (load)	kN/m	0.06853
6.895	psi (stress)	kPa	0.1450
6.895	ksi (stress)	MPa	0.1450
0.04788	psf (load or pressure)	kPa	20.93
47.88	ksf (load or pressure)	kPa	0.02093
16.02	pcf (density)	kg/m^3	0.06242
$0.566 \times (°F - 32)$	°F	°C	$(1.8 \times °C) + 32$

corresponding units in the SI system. The conversion factors used in shifting from one system to the other are given in Table 1.3.

1.8. STANDARD SYMBOLS

The following "shorthand" symbols are frequently used.

Symbol	Reading
>	is greater than
<	is less than
≥	equal to or greater than
≤	equal to or less than
6 ft	6 feet
6 in.	6 inches
Σ	the sum of
ΔL	change in L

1.9 NOTATION

Use of standard notation is complicated by the fact that there is some lack of consistency in the notation used in the various fields of design that overlap in site design work (architecture, civil engineering, geology, foundation engineering, etc.). To keep some form of consistency in this book, the following nota-

tion has been used, most of which is in general agreement with that used in basic engineering work.

a	Increment of an area (sq ft, sq in., etc.)
A	Gross area (sq ft, sq in., etc.)
D	Diameter
e	Eccentricity, as a dimension of mislocation of something
f	computed unit stress (psi, psf, etc.)
F	(1) Force; (2) allowable unit stress
h	Height, as a measured distance
H	Horizontal component of a force
l	Length dimension
L	Length dimension
N	Number of
p	unit pressure, as compressive stress or force
P	Concentrated load (force at a point)
R	Radius
s	Spacing dimension, usually center-to-center of a set of objects
t	Thickness dimension
T	Temperature
w	(1) Width dimension; (2) unit weight
W	Gross weight
Δ (delta)	Change of
θ (theta)	Angle
Σ (sigma)	Sum of
ϕ (phi)	Angle

2

MATHEMATICS FOR SITE ENGINEERING

This chapter presents explanations of the use of some basic mathematics with direct applications to work in site engineering. This work is not intended to replace a more thoroughly developed treatment in a course in mathematics, but should provide some assistance for readers with limited backgrounds in mathematical training. The work here consists of some elementary topics in geometry and trigonometry. For readers with extensive preparation in mathematics, this work may not be necessary, although it may still be useful as a refresher and as an indicator of direct applications to the work in this book.

2.1 GRAPHIC SOLUTIONS OF TRIANGLES

The solution of problems relating to triangles may be performed by drawing to scale the known sides and angles and scaling the unknown parts. The results found by this graphical method are not sufficiently accurate for

most purposes, but it is advantageous to keep the method in mind. Frequently, such a drawing, when used as a check, reveals errors that have occurred in the mathematical computations.

2.2 THE RIGHT TRIANGLE

Figure 2.1a shows a right triangle with the conventional lettering that identifies the various parts. The three sides are a, b, and c (the hypothenuse). The right angle is angle C, angle A is the interior angle between sides c and b, and angle B is the interior angle between sides c and a. The right triangle and the relations of its sides and angles form the basis of plane trigonometry. Whereas many problems in trigonometry are complex and involved, the ability to solve problems relating to right triangles enables one to perform many computations in connection with surveying that are commonly met with in practice.

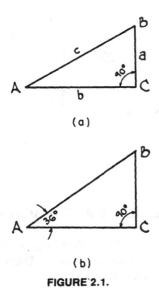

(a)

(b)

FIGURE 2.1.

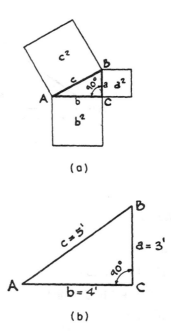

(a)

(b)

FIGURE 2.2.

2.3 GEOMETRIC PRINCIPLES

Two important principles found in the study of geometry are of great assistance in the solution of triangles.

First. The sum of the interior angles of a triangle is equal to 180°. In a right triangle one of the angles is a 90° angle. Consequently, the sum of the remaining two acute angles is 90°. Therefore, if one acute angle is known, this angle subtracted from 90° determines the magnitude of the third angle. As an example, consider the right triangle shown in Figure 2.1b. Angle C is 90° and angle A is 36°. To find angle B, we simply subtract 36° from 90°. Thus, angle $B = 90° - 36°$, or 54°, and $A + B + C = 36 + 54 + 90 = 180°$.

Second. In any right triangle the square of the hypothenuse is equal to the sum of the squares of the other two sides. This is known as the Pythagorean theorem. The hypothenuse is side c, the side opposite the right angle, as shown in Figure 2.1a. It is always the longest side. Referring to Figure 2.2a, $c^2 = a^2 + b^2$. This important principle may be used to determine the unknown side of a right

triangle when the remaining two sides are known. Consider, for example, the right triangle shown in Figure 2.2b. Side c, the hypothenuse, is 5 in. in length, and side a has a length of 3 in. Determine the length of b, the remaining side. In accordance with the above principle,

$$c^2 = a^2 + b^2 \qquad \text{or } 5^2 = 3^2 + b^2$$

Then

$$b^2 = 25 - 9 = 16 \qquad \text{or } b$$
$$= \sqrt{16} \text{ and } b = 4 \text{ ft}$$

The right triangle whose sides have 3, 4, and 5 units of length is sometimes called "the magic triangle." Without the use of a surveying instrument, builders frequently lay out right angles with tapes, using triangles in this proportion: 3, 4, and 5; 15, 20, and 25, etc.

Example 1. Figure 2.3a represents a right triangle in which side $a = 17.62$ ft and side $b = 23.21$ ft. Determine the length of side c, the hypothenuse.

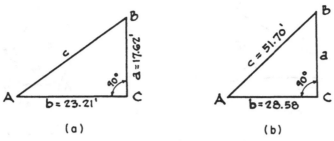

FIGURE 2.3.

Solution:

$$c^2 = a^2 + b^2$$

$$c^2 = 23.21^2 + 17.62^2$$

$$c^2 = 538.7 + 310.5 = 849.2$$

$$c = \sqrt{849.2}$$

$c = 29.14$ ft the length of the hypothenuse

Example 2. Figure 2.3*b* shows a right triangle in which $c = 51.70$ ft and $b = 28.56$ ft. Find the length of side *a*.

PROBLEMS 2.3.A THROUGH F.

For the following right triangles the lengths of certain sides are given. Determine the length of the unknown side.

Problem	Given	Find
2.3.A	$a = 61.17$ ft and $b = 32.78$ ft	c
2.3.B	$a = 4.50$ ft and $b = 8.12$ ft	c
2.3.C	$a = 52.27$ ft and $c = 63.08$ ft	b
2.3.D	$a = 14.06$ ft and $c = 14.32$ ft	b
2.3.E	$b = 10.00$ ft and $c = 14.14$ ft	a
2.3.F	$b = 59.23$ ft and $c = 72.76$ ft	a

Solution:

$$c^2 = a^2 + b^2$$

$$51.70^2 = a^2 + 28.58^2 \text{ hence } a^2 = 51.70^2 - 28.58^2$$

$$a^2 = 2673 - 816.8 = 1856.2, \text{ say } 1856$$

$$a = \sqrt{1856}$$

$$a = 43.08 \text{ ft}$$

For the following right triangles, verify the lengths of the unknown sides.

Given	Find	Answer
$a = 21.36$ ft and $b = 60.52$ ft	c	$c = 64.18$ ft
$a = 41.23$ ft and $b = 13.50$ ft	c	$c = 43.38$ ft
$a = 76.10$ ft and $c = 82.31$ ft	b	$b = 31.37$ ft
$a = 8.36$ ft and $c = 96.75$ ft	b	$b = 96.39$ ft
$b = 26.28$ ft and $c = 35.98$ ft	a	$a = 24.58$ ft
$b = 5.23$ ft and $c = 5.33$ ft	a	$a = 1.03$ ft

2.4 TRIGONOMETRIC FUNCTIONS OF ANGLES

Figure 2.4a shows a right triangle in which angle A is 30°. The lengths of sides a, b, and c are 1, $\sqrt{3}$, and 2, respectively. In any right triangle in which A is 30°, no matter how large or how small the triangle, the ratio of the side opposite angle A to the hypothenuse, a/c, will always be $\frac{1}{2}$; the ratio of the adjacent side to the hypothenuse, b/c, will always be $\sqrt{3}/2$; the ratio of the side opposite angle A to the adjacent side, a/b, will always be $1/\sqrt{3}$, and so on. The ratio of the length of one side to the length of another side is known as a *trigonometric function* of the angle in question. These ratios have specific names, *sine*, *cosine*, *tangent*, etc.; but keep in mind that *they are simply ratios*.

In any right triangle there are three sides and two acute angles, an acute angle being an angle between 0° and 90°. Consequently, there are six ratios (functions); they depend on the size of the angle regardless of the size of the triangle. Refer to Figure 2.4b. The names given to the functions are as follows:

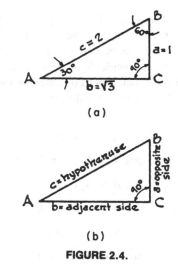

(a)

(b)

FIGURE 2.4.

The *sine*, *cosine*, and *tangent* are the functions used most frequently, and these ratios, a/c, b/c, and a/b, should be memorized. Note particularly that *the preceding ratios are functions of angle* A.

The other acute angle in the right triangle, Figure 2.4b, is angle B. Now let us consider the functions of angle B. Since, from the

$$\textit{sine of angle } A = \frac{\text{opposite side}}{\text{hypothenuse}} = \frac{a}{c} \qquad \text{Abbrev.} = \sin A$$

$$\textit{cosine of angle } A = \frac{\text{adjacent side}}{\text{hypothenuse}} = \frac{b}{c} \qquad \text{Abbrev.} = \cos A$$

$$\textit{tangent of angle } A = \frac{\text{opposite side}}{\text{adjacent side}} = \frac{a}{b} \qquad \text{Abbrev.} = \tan A$$

$$\textit{cotangent of angle } A = \frac{\text{adjacent side}}{\text{opposite side}} = \frac{b}{a} \qquad \text{Abbrev.} = \cot A$$

$$\textit{secant of angle } A = \frac{\text{hypothenuse}}{\text{adjacent side}} = \frac{c}{b} \qquad \text{Abbrev.} = \sec A$$

$$\textit{cosecant of angle } A = \frac{\text{hypothenuse}}{\text{opposite side}} = \frac{c}{a} \qquad \text{Abbrev.} = \csc A$$

above, the sine of an angle is opposite side/hypothenuse, the sine of angle $B = b/c$. Similarly, the cosine of angle $B = a/c$ and the tangent of angle $B = b/a$. Referring to the conventional system of lettering shown in Figure 2.4b, note the following relationships:

$$\sin A = a/c \cos B$$

$$\cos A = b/c \sin B$$

$$\tan A = a/b \cot B$$

$$\sec A = c/b \csc B$$

$$\csc A = c/a \sec B$$

$$\cot A = b/a \tan B$$

In the right triangle shown in Figure 2.4b, angles A and B are acute angles and angle A + angle $B = 90°$. Either angle is called the *complement* of the other. For instance, the complement of $40°$ is $90° - 40°$ or $50°$. In Figure 2.4b $\sin A = a/c$ and $\cos B = a/c$; that is, $\sin A = \cos B$. Consider the following pairs of functions: sine and cosine, tangent and cotangent, secant and cosecant. In each pair either function is the *cofunction* of the other one. Any function of angle A equals the cofunction of angle B. Thus, $\sin 30° = \cos 60°$, $\tan 26°40' = \cot 63°20'$, and $\sec 50° = \csc 40°$.

Two right triangles are much used by architects. In one the acute angles are each $45°$ (see Figure 2.5a), and in the other they are $30°$ and $60°$, as shown in Figure 2.4a. Side a is one unit of length, and the lengths of the other sides are shown in the figures. From these two figures the data in Table 2.1 are readily compiled.

Most of the trigonometric functions shown in the tabulation are fractions. Expressed as decimals, $1/2 = 0.5$, $\sqrt{3}/2 = 0.86603$, $1/\sqrt{3} = 0.57735$, $\sqrt{3} = 1.73205$, $2/\sqrt{3} = 1.15470$, $1/\sqrt{2} = 0.70711$ and $\sqrt{2} = 1.41421$. Thus, we know that $\sin 30° = 0.5$, $\cos 30° = 0.86603$, $\tan 60° = 1.73205$, $\cos 45° = 0.70711$, etc. These values are called

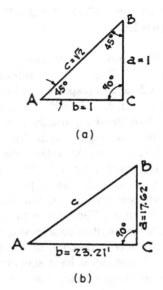

(a)

(b)

FIGURE 2.5.

the *natural trigonometric functions*. By referring to a table of natural trigonometric functions we may find directly the function of any angle between $0°$ and $90°$.

2.5 FINDING TWO ANGLES WHEN TWO SIDES ARE KNOWN

As explained in Section 2.3, the third side of a right triangle may be found if the lengths of the other two sides are given. Now that we are familiar with the various trigonometric functions explained in Section 2.4, we may find the two acute angles when the lengths of two sides are known.

Example. Figure 2.5b shows a right triangle in which the lengths of two sides are given. Determine the two acute angles A and B.

Solution: Since $\tan A$ involves sides a and b, both of which are known, we may write

$$\tan A = \frac{a}{b} = \frac{17.62}{23.21}$$

TABLE 2.1   TRIGONOMETRIC FUNCTIONS OF COMMON ANGLES

Angle	Sin	Cos	Tan	Cot	Sec	Csc
30°	$\frac{1}{2}$	$\frac{\sqrt{3}}{2}$	$\frac{1}{\sqrt{3}}$	$\sqrt{3}$	$\frac{2}{\sqrt{3}}$	2
60°	$\frac{\sqrt{3}}{2}$	$\frac{1}{2}$	$\sqrt{3}$	$\frac{1}{\sqrt{3}}$	2	$\frac{2}{\sqrt{3}}$
45°	$\frac{1}{\sqrt{2}}$	$\frac{1}{\sqrt{2}}$	1	1	$\sqrt{2}$	$\sqrt{2}$

Thus, angle A is the angle whose tangent is 17.62/23.21 (called the *arctangent* or *arctan* of the number). Using a calculator or a table of trigonometric functions, we determine

$$A = \arctan \frac{17.62}{23.21} = 37.204°$$

Similarly,

$$B = \arctan \frac{23.21}{17.62} = 52.796°$$

For a check, we may observe that the sum of $A + B$ should be 90°, which is the case.

2.6   CHECKING COMPUTATIONS

In the solutions of triangles as illustrated in Section 2.5, there should be no doubt about the correctness of the result for it is always possible to use a different method of solution in order to check the answer. For the problem given in Section 2.5, angle A was determined first; it was found to be 37.204°.

Since

$$\text{angle } A + \text{angle } B = 90°$$
$$\text{angle } B = 90° - \text{angle } A$$

or

$$\text{angle } B = 90° - 37.204°$$

and

$$\text{angle } B = 52.796°$$

But this method is no check on the computations involved in solving angle A, for in determining angle B it assumes that angle A is correct. Determining the two angles by different methods and then adding them, knowing that their sum should be 90°0′, is a valid check on the computations. This procedure was followed in the example. A check should always be made.

For the right triangles in Table 2.2, the

TABLE 2.2   COMPUTATION OF ANGLES

Given Side Dimensions (ft)		Angles (degrees)	
		A	B
a = 4.72	b = 12.26	21.06	68.94
a = 81.73	b = 44.19	61.60	28.40
a = 7.06	c = 11.17	39.20	50.80
a = 15.92	c = 52.25	17.74	72.26
b = 19.98	c = 46.23	64.39	25.61
b = 38.26	c = 54.01	44.90	45.10

lengths of two sides are given and the two acute angles have been determined by computations. Verify the sizes of the angles.

For the triangles described in Table 2.2 observe that the function of angle A is always the cofunction of angle B; that is, $\sin A = \cos B$, $\tan A = \cot B$, and so on.

PROBLEMS 2.6.A THROUGH F.

For the following right triangles, described by the lengths of two sides, find the sizes of the two acute angles.

Problem	Given
2.6.A	$a = 27.21$ ft and $b = 53.08$ ft
2.6.B	$a = 3.01$ ft and $b = 4.67$ ft
2.6.C	$a = 10.09$ ft and $c = 74.99$ ft
2.6.D	$a = 3.08$ ft and $c = 4.28$ ft
2.6.E	$b = 61.05$ ft and $c = 69.99$ ft
2.6.F	$b = 6.69$ ft and $c = 35.58$ ft

2.7 SOLVING RIGHT TRIANGLES WITH ONE SIDE AND ONE ACUTE ANGLE KNOWN

When one side and an acute angle of a right triangle are known, the lengths of the remaining sides may be computed by the use of the trigonometric functions and the logarithmic tables. The other acute angle is found by subtracting the known angle from 90°.

Example. In Figure 2.6a angle A is 33.33° and side b has a length of 52.33 ft. Find the remaining parts of the right triangle.

Solution: Since the sum of the two acute angles is 90°, 33.33° + angle B = 90°, angle B = 90° − 33.33°, and angle B = 56.67°.

The tangent of angle A involves sides a and b, (Section 2.4). Since side b is known, we can compute the length of side a.

$$\tan A = \frac{a}{b}$$

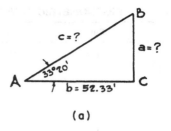

(a)

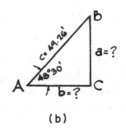

(b)

FIGURE 2.6.

Substituting, $\tan 33.33° = \dfrac{a}{52.33}$ and $a = 52.33 \times \tan 33.33°$.

From which, side $a = 34.42$ ft

Since sides b and c are included in the cosine of angle A, $\cos A$ enables us to compute the length of side c. Thus,

$$\cos A = \frac{b}{c}$$

Substituting, $\cos 33.33° = \dfrac{52.33}{c}$ and $c = \dfrac{52.33}{\cos 33.33°}$.

From which, side $c = 62.63$ ft.

After a was computed we might have determined c by using the sine angle A. Thus, $\sin A = a/c$. By this formula we could establish c, but it would be based on the assumption that no error had been made in computing a. If a mistake had been made in computing a, c would also have been in error. The method used above in determining c is the proper method for it does not involve the computed side a. *Make it a rule, in determin-*

ing unknown parts, to use only values given as data.

By data, we were given the length of b and, by computations, we have determined the lengths of a and c. Now, to check the correctness of these lengths we can apply the principle given in Section 2.3. Thus,

$$c^2 = a^2 + b^2$$

Substituting,

$$62.63^2 = 34.42^2 + 52.33^2$$
$$3923 = 1185 + 2738$$

which provides the check on the computations.

2.8 ARRANGEMENT OF COMPUTATIONS

It is of importance that pains be taken to arrange all computations in a neat, legible, and systematic manner. A habit thus formed will result in a saving of time and will minimize the possibility of error. The practice of checking results will be found to be of great value. This is particularly true in engineering work because the solution of one problem frequently provides the data for ensuing problems. Thus, an error at the beginning, if not detected, may result in wasted effort and confusion. Whenever possible, diagrams should be drawn to scale. Such diagrams frequently reveal the presence of errors.

The following example is presented, without explanatory notes, as a suggested form of computation arrangement.

Example. In the right triangle shown in Figure 2.6*b*, angle $A = 48.5°$ and $c = 49.26$ ft. Determine angle B and the lengths of sides a and b.

Solution:

$$A + B = 90°$$

Therefore,

$$48.5 + B = 90°, \; B = 41.5°$$

To find side a,

$$\sin 48.5° = \frac{a}{49.26}, \; a = 49.26 \, (\sin 48.5°)$$

From which, side $a = 36.89$ ft
To find side b,

$$\cos 48.5° = \frac{b}{49.26}, \; b = 49.26 \, (\cos 48.5°)$$

From which, side $b = 32.64$ ft
Check:

$$49.26^2 = 36.89^2 + 32.64^2$$
$$2427 = 1361 + 1065$$

which provides a reasonable check.

Table 2.3 displays data for several problems similar to the preceding example. Verify the computed data for each table entry, using the procedure just demonstrated.

TABLE 2.3 COMPUTED VALUES FOR TRIANGLES

Given	Computed
$A = 30°$, $c = 28.06$ ft	$B = 60°$, $a = 14.03$ ft, $b = 24.30$ ft
$B = 22.5°$, $c = 73.26$ ft	$A = 67.5°$, $a = 67.68$ ft, b 28.04 ft
$A = 17.93°$, $a = 12.68$ ft	$B = 72.07°$, $b = 39.18$ ft, $c = 41.18$ ft
$B = 46.38°$, $b = 56.73$ ft	$A = 43.62°$, $a = 54.05$ ft, $c = 78.36$ ft
$A = 72.68°$, $b = 8.23$ ft	$A = 17.32°$, $a = 26.40$ ft, $b = 27.65$ ft
$B = 62.1°$, $a = 31.17$ ft	$A = 27.9°$, $b = 58.87$ ft, $c = 66.61$ ft

PROBLEMS 2.8.A THROUGH F.

For each of the following right triangles one acute angle and one side are given. Find the unknown parts.

Problem	Given	Find
2.8.A	$A = 45°$, $c = 14.14$ ft	B, a, and b
2.8.B	$B = 81.017°$, $c = 92.32$ ft	A, a, and b
2.8.C	$A = 36.7°$, $a = 6.04$ ft	B, b, and c
2.8.D	$B = 15.25°$, $b = 11.12$ ft	A, a, and c
2.8.E	$A = 8.27°$, $b = 34.23$ ft	B, a, and c
2.8.F	$B = 45.083°$, $a = 26.01$ ft	A, b, and c

2.9 OBLIQUE TRIANGLES AND THE SINE LAW

Whereas the principles involved in the solution of right triangles may be used in solving any triangle, many trigonometric formulas have been derived which simplify the solution of certain problems. Among these formulas is the *sine law*.

The Sine Law. In any triangle the sides are proportional to the sines of the opposite angles. Figure 2.7a shows any triangle and, in accordance with the sine law,

$$\frac{a}{\sin A} = \frac{b}{\sin B} = \frac{c}{\sin C}$$

In Section 2.4, referring to Figure 2.4b, a right triangle, we found that

$$\sin A = \frac{\text{opposite side}}{\text{hypothenuse}} = \frac{a}{c}$$

Figure 2.7b shows angle A to be greater than 90°, an *obtuse angle*. From the figure we see that $\sin A = a/c$. The numerical value of sin $A = \sin (180 - A)$.

Example. In Figure 2.7a, angle $A = 105°$, and angle $B = 30°$, and angle $C = 45°$. If side $a = 50$ ft, find the lengths of sides b and c.

Solution: Since angle A is an obtuse angle, sin $A = \sin (180° - 105°)$, or sin $A = \sin 75°$. Now, applying the sine law,

$$\frac{a}{\sin A} = \frac{b}{\sin B} \text{ or } b = \frac{50 \sin 30°}{\sin 75°}$$

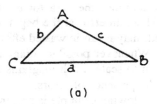

(a)

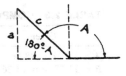

(b)

FIGURE 2.7.

From which, $b = 25.88$ ft

Similarly,

$$\frac{a}{\sin A} = \frac{c}{\sin C} \text{ or } c = \frac{50 \sin 45°}{\sin 75°}$$

From which, $c = 36.60$ ft

PROBLEM 2.9.A.

For the triangle shown in Figure 2.8a, find the lengths of sides AB and BC and also angle A.

PROBLEM 2.9.B.

For the triangle shown in Figure 2.8b, find the lengths of sides AB and BC and also angle B.

2.10 AREAS OF TRIANGLES

The area of a right triangle may be found by taking $\frac{1}{2}$ the product of the base by the height.

When the lengths of all three sides of a triangle are known, the area of any triangle may be found by use of the following formula:

$$\text{Area} = \sqrt{s(s - a)(s - b)(s - c)}$$

in which a, b, and c are the lengths of the sides of the triangle and $s = \frac{1}{2}(a + b + c)$.

Example. Compute the area of the triangle shown in Figure 2.9a.

Solution: First, let us compute the magnitude of s.

$$s = \frac{30.17 + 20.20 + 42.83}{2} = \frac{93.20}{2}$$

$$= 46.60$$

Then,

$$\begin{array}{r} 46.60 \\ -30.17 \\ \hline (s - a) = 16.43 \end{array}$$

$$\begin{array}{r} 46.60 \\ -20.20 \\ \hline (s - b) = 26.40 \end{array}$$

$$\begin{array}{r} 46.60 \\ -42.83 \\ \hline (s - c) = 3.77 \end{array}$$

$$\text{Area} = \sqrt{46.60 \times 16.43 \times 26.40 \times 3.77}$$

$$= 276.1 \text{ ft}^2$$

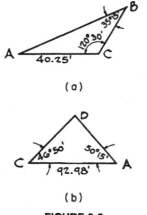

(a)

(b)

FIGURE 2.8.

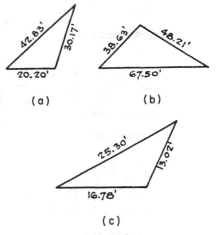

(a)

(b)

(c)

FIGURE 2.9.

PROBLEM 2.10.A.

Compute the area of the triangle shown in Figure 2.9b.

PROBLEM 2.10.B.

Compute the area of the triangle shown in Figure 2.9c.

2.11 PROPERTIES OF COMMON GEOMETRIC FORMS

For various purposes in site design work it is necessary to compute quantities for geometric forms. Examples of such computations are those required to find the total linear length for a property line fence, the surface area of a portion of open ground for planting, and the volume of an excavation. Such elements may

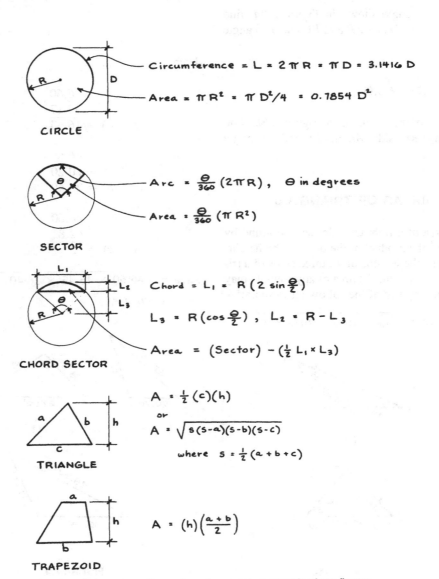

CIRCLE

Circumference $= L = 2\pi R = \pi D = 3.1416\,D$

Area $= \pi R^2 = \pi D^2/4 = 0.7854\,D^2$

SECTOR

Arc $= \dfrac{\theta}{360}(2\pi R)$, θ in degrees

Area $= \dfrac{\theta}{360}(\pi R^2)$

CHORD SECTOR

Chord $= L_1 = R\left(2\sin\dfrac{\theta}{2}\right)$

$L_3 = R\left(\cos\dfrac{\theta}{2}\right)$, $L_2 = R - L_3$

Area $= (\text{Sector}) - \left(\dfrac{1}{2}L_1 \times L_3\right)$

TRIANGLE

$A = \dfrac{1}{2}(c)(h)$

or

$A = \sqrt{s(s-a)(s-b)(s-c)}$

where $s = \dfrac{1}{2}(a+b+c)$

TRAPEZOID

$A = (h)\left(\dfrac{a+b}{2}\right)$

FIGURE 2.10. Properties of common geometric plane figures.

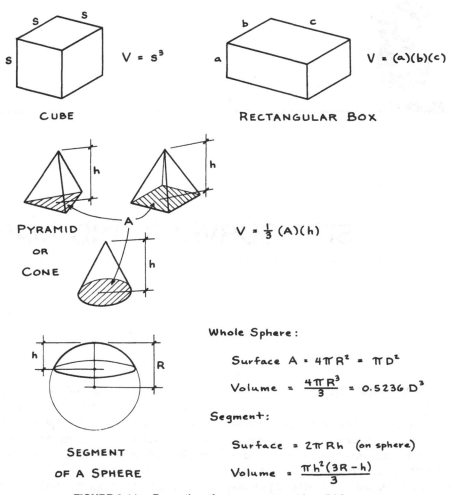

FIGURE 2.11. Properties of common geometric solid forms.

be considerably irregular in shape, but often consist simply of ordinary geometric forms; or at the most, of multiples of simple forms.

Even very irregular forms may often be broken down for reasonable approximation into multiples of simple forms. This may be reasonably adequate for preliminary estimates or even for final work. Use of presently available computer-aided procedures makes this less useful for final design work, but simple hand computations for preliminary work are still valuable.

Various tasks in surveying involving geometric forms are described throughout this book. The purpose here is to provide data for determination of some of the most elementary elements that commonly exist or which can be used to approximate more complex, irregular forms.

Figure 2.10 presents a number of common planar geometric forms with data for determination of lengths and areas. Figure 2.11 presents some common three-dimensional shapes with data for determination of surface areas and volumes.

3

SITE SURVEYS AND MAPS

3.1 SITE INFORMATION

Whenever the development of a site is planned some information about the site is required. The type and extent of information vary, depending both on site conditions and the type of planned development. The general process of gathering information is sometimes described as *surveying* the sight, although the actual work of surveying is itself a special process of finding specific dimensions—both horizontal and vertical. Information determined by surveying and other investigations is often displayed on a map (or plan) of the site.

3.2 TYPES OF SITE SURVEYS

The term *site survey* is usually reserved for a special map that is produced by a registered *land surveyor* and is registered with local authorities as a legal document that becomes part of the legal description of the property. However, the specific legal definition of a site is contained in a written description of the property that refers to its location with respect to established legal boundaries of the region (city, county, etc.).

In addition to the site boundaries, a survey indicates various other information, such as

Locations of adjacent streets and alleys.

Easements for utilities (portions of the property that have existing utilities or are held available for future installations).

Locations and descriptions of major site features, such as ponds, streams, rock out-croppings, existing buildings, large trees, and so on.

The site survey is a general description of the site, with an emphasis on surface features. Other information about a site may be use-

20

ful—or specifically required—for the planning of site development. Additional types, sources, and uses of information include the following.

General Area Map These may be obtained from local authorities, from various agencies (highway, agriculture, U.S. Army, etc.), or in some cases from commercial mapping services.

Geographic Statistics These are maps displaying distribution of various data, such as population density, air pollution, seismicity, snow or rainfall, degree days, and so on.

Aerial Surveys (Photo Maps) These exist for many regions of the United States and are generally obtainable from government agencies or commercial mapping services.

Geotechnical Surveys These deal with information on ground conditions and geological properties of the surface and subsurface ground materials. Some existing information may be available from general surveys made by agencies or from studies for previous site development. Obtaining a permit for any site development usually requires the performing of such a survey prior to any construction activity.

In any area where considerable development already exists, a great deal of information typically exists for the general conditions of the region of a site. This information can be used, together with a visit to the site, to get a preview of site conditions.

The exact type and extent of new information required for the planning and development of a site will depend on the nature of the planned work, the specific requirements of local government agencies, unique site conditions, and the amount of existing information.

3.3 SITE DEVELOPMENT PLANS

Geographical information, legal property descriptions, and site surveys of various kinds are generally related to the production of maps. Maps are essentially horizontal, planar views of some region of the earth's surface. Data may be recorded directly on the map or refer to some identified portion of the map.

The making of maps (called *cartography*) is a highly developed science that uses procedures and forms of data that are quite specifically established. Any references to internationally-established latitudes or longitudes, to boundaries of countries, states, counties, or municipalities, or to legally defined private properties must follow carefully defined procedures.

In addition, forms of information such as locations of streets and highways, established survey references (such as benchmarks, See Section 8.4), height restrictions for construction, or property easements must be displayed in established formats.

In general, maps are produced and used in highly controlled situations. Plans, on the other hand, are developed for specific purposes, and the form, as well as the type of information displayed, relate to the type of activity for which the plans are prepared. For site development, some typical types of plans frequently used are the following.

Site Plan This is essentially a site map, typically created for some actual mapping purposes that result in repetition of data from site surveys, but also displaying various aspects of the proposed work. For simple projects a single site plan may suffice. For large, complex projects, a series of plans may display selected information for greater ease of readability. The site plans are typically part of the set of construction drawings for a construction project.

Grading Plan This is generally a site map that displays the existing site contours (form of the surface) and existing features (such as trees, existing construction, etc.), and also indicates the form of the recontoured and generally redeveloped site surface. It is prepared specifically to inform the workers responsible for recontouring the site surface.

Construction Plans These are plans showing the horizontal planar view of the construction work proposed for a site. The construction itself will typically be additionally described with various detailed drawings. Separate plans may be drawn for site construction (drives, curbs, retaining structures, planters, etc.), building foundations, other below-grade construction (basements, tunnels, etc.), and the grade level (first floor) of proposed buildings.

3.4   DATA SOURCES

Maps are generally used primarily as data sources to inform the process of design for site and building development. They describe an "as is" condition, from which plans may be developed for proposed work. Within the site boundaries, changes can be made with some freedom, although typically limited by many practical restrictions as well as some legal ones.

Site development is typically constrained by many aspects of the boundary conditions. The simple dimensions of boundaries must be recognized, resulting in a limit of the the extent of horizontal projection of the construction. However, many other aspects of planning must relate to recognition of boundaries and adjacent features. Some specific concerns are the following.

Surface Drainage Recontouring of the site, as well as new construction, must not cause problems for adjacent properties in terms of rechannelling of surface water drainage during rainfalls.

Existing Streets, etc. These usually represent unchangeable conditions that must be dealt with in recontouring, planning drives, and so on. Vehicular entry and exit on the site must recognize traffic conditions and other restrictions related to existing elements.

Existing Utilities Connections to existing services (for power, water, gas, phones, etc.) must usually be made with recognition of existing mains. Of especially critical concern are sewers, which work by gravity drainage, making vertical location of on-site elements critical.

Adjacent Properties Construction on the site must not jeopardize adjacent properties by presenting dangers of undermining, erosion, etc.

A major purpose of the surveys is to establish all the data necessary for an informed design that recognizes all of these concerns, as well as others appropriate to the proposed work or adjacent conditions.

3.5   DESIGN DEVELOPMENT

Design of construction projects typically proceeds in a somewhat staggered fashion. At the earliest stages, very broad decisions must be made without much detailed information. Detailed information must usually be obtained on the basis of some actual design studies (preliminary design developments), so that the purposes for the detailed information (such as geotechnical surveys) are somewhat specifically determined.

Thus, the generation of the information that supports the design work, and the design work itself, must be developed in a step-by-step procedure. First, some general information, then some preliminary design, then some more specific information, then some more definitive design, then some very critical, special information, and so on. This process may actually continue into and through the construction work, since some information may only be obtainable during the work of redeveloping the site. (What is truly down there below the ground surface at some specific point on the site?)

What is important is to anticipate the need for information at various stages of design, to recognize the feasibility of obtaining various forms of information, and to plan for the general flow and interchange of design work and information gathering. In the best of situations, design work will not proceed uninformed and information will be obtained in a proper and timely fashion.

4

MEASURING DISTANCES

A basic task in surveying work is that of the simple linear measurement of distances. This work can now largely be achieved with surveying equipment, but for many situations it is still practical to use simpler methods, the most common being the use of a marked tape. This chapter explains the use of the tape for ordinary measurements.

4.1 TAPES

All tapes are subject to stretching when tension is applied and, because of its greater resistance to deformation, steel tape is used exclusively for accurate measurements. The steel tape is commonly available in 50 ft and 100 ft lengths and in greater lengths when required. It is a thin ribbon of steel graduated in feet and fractions of a foot. Generally the foot is divided into tenths and hundredths. Before using the tape, observe carefully the position of the zero point. Sometimes this

point is at the extreme end of the ring at the end of the tape, as shown in Figure 4.1a. Sometimes, however, as indicated in Figure 4.1b, it is marked on the body of the tape. The advantage in using the former is that one may hook the end over a pin in the ground or a nail in the wall without the necessity of having someone hold the end of the tape. Much time may be saved and greater accuracy obtained by measuring distances with the aid of an assistant.

4.2 CHAINS

Many old surveys have distances indicated in units of *chains*. A chain consisted of 100 links of heavy steel wire. Each link had a ring at its ends by means of which it was joined to an adjacent link. A chain was 66 ft in length and, since there were 100 links in a chain, each link had a length of 0.66 ft or 7.92 in. Ten square chains equaled 1 acre and, conse-

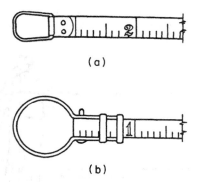

(a)

(b)

FIGURE 4.1. Forms of tapes.

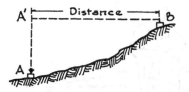

FIGURE 4.2.

quently, the use of a chain afforded a convenient measure in surveying land acreage.

By constant use, the wear on the rings of a chain resulted in its lengthening and a subsequent error in dimensions. This is thought to be the reason for the discrepancies between the *District Standards* of length, used in some of our older localities, and the *U.S. Standard* of measurement. The chain is no longer used as a unit of measure in this country. In converting chains to feet and inches in old deeds or surveys, the following conversion table may be used:

7.92 in. = 1 link

100 links = 1 chain = 66 ft

80 chains = 1 mile

4.3 HORIZONTAL DISTANCES

It is of the utmost importance to bear in mind that *all distances shown on maps and plans are horizontal projections of distances*. The surface of the ground generally has a slope and seldom lies in a horizontal plane. However, the distances recorded on surveys or in deed descriptions are always horizontal distances. Figure 4.2 shows a cross section taken through the side of a hill. The distance between points *A* and *B* would be measured and recorded as the distance between *A'* and *B*,

actually the horizontal projection of the distance between points *A* and *B*.

Generally, distances are measured by holding the tape in a horizontal position and transferring the distance between two points by means of a plumb line. One man holds the zero end of the tape over a tack on a stake *at the higher level* while the other man, holding the tape level with one hand and with a plumb line in the other, places the plumb bob directly over the point to be measured and reads the distance at which the plumb line intersects the tape. This procedure is indicated in Figure 4.3. Experiment will show that *it is much easier to measure downhill than uphill*.

From the figure, it is obvious that the two ends of the tape must be held at the same level; the tape must be horizontal if the measurement is to be accurate. Naturally, there will be a certain degree of sag in the tape. It is customary to maintain about a 10-lb tension to measure with a sufficient degree of accuracy for ordinary work. For precise city surveying, a level is sometimes placed on the tape and a spring balance employed to maintain a uniform tension.

Measurements may be made by laying the tape directly on the slope. Suppose, for example, we wish to determine the horizontal distance *A'B* between the two points *A* and *B* shown in Figure 4.4a. The tape is laid on the surface of the slope and the dimension *AB* is determined. Then the angle *θ* is measured by a transit and, by the principles of trigonometry given in Chapter 2, the distance *A'B* is computed. This method of measuring requires considerable time and is used only when required by conditions in the field.

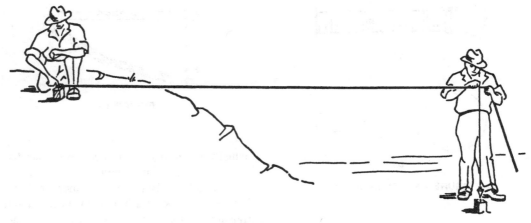

FIGURE 4.3.

4.4 MEASURING ON STEEP SLOPES

For steep slopes it may be impracticable to hold the tape level for the entire distance. For such a condition, the distance between the two points to be measured is divided into a number of parts so that the forward man may hold the tape conveniently at chest height.

Referring to Figure 4.4b, suppose it is required to measure the horizontal distance between points A and B, the difference in the elevation between the two points being approximately 10 ft.

We begin by having the rear tapeman hold the zero point of the tape over the upper stake, point A. The head tapeman goes forward and the rear tapeman directs him to move to the right or left until he is on the line between points A and B. At a convenient point at some even foot mark, as at M, he places a pin in

the ground and calls this dimension, say 55 ft, to the rear tapeman. The rear tapeman should repeat this figure aloud to avoid error. Then the rear tapeman drops the tape, advances to M, and holds the 55 ft mark on the tape over the pin. The forward tapeman stretches the tape to its full length, say 100 ft, and places a pin in the ground at point N. The rear tapeman advances with the tape and holds the zero point over this pin. Then the head tapeman proceeds forward, plumbs over point B, and reads the tape; the dimension is 63.18 ft. Thus the horizontal distance between points A and B is 163.18 ft.

4.5 ALIGNMENT BETWEEN POINTS

When the distance to be measured between two points must be measured in parts, or when it exceeds the length of the tape, it is impor-

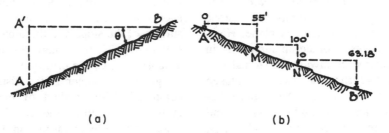

(a)   (b)

FIGURE 4.4.

FIGURE 4.5.

tant that the intervening points be on an approximately straight line. By inspection of Figure 4.5, it is seen that if the points M and N are not on the line between points A and B (shown in plan) the recorded distance, $AM + MN + NB$, will be greater than the true distance AB. In order to prevent this inaccuracy, the rear tapeman sights from A to B and directs the head tapeman to the right or left in establishing the intervening points. In cases in which the termination point is not visible to the rear tapeman, as, for instance, an obstructing knoll, a *ranging pole* is placed in the ground close to the termination point and the rear tapeman sights on this. A ranging pole is usually of wood, octagonal in cross section, having lengths of 6 ft to 10 ft. The pole is enameled red and white in alternate bands and the end is fitted with a steel point.

4.6 MARKING POINTS ON A SURVEY

The ends of straight lines on a survey are generally marked by driving wooden stakes into the ground, the exact point being located by a tack or brad driven in the stake. When it is not feasible to drive wooden stakes, metal spikes may be used, the exact point being marked by a center punch in the head of the spike. In rock, concrete pavements, or masonry portions of buildings, a yellow chalk mark is often used, the exact point being located with a black pencil cross on the chalk. For permanent locations, the point may be marked by a small drilled hole.

For temporary locations, and for intermediate points, metal pins, called arrows, are commonly used. These metal pins are approximately $\frac{3}{16}$ in. in diameter and 12 in. to 15 in. in length. They are driven into the ground inclined to the vertical and at right angles to

the line between the points being measured. The exact point being marked is the center of the pin at the surface of the ground.

4.7 MISTAKES AND ERRORS

A *mistake* is the result of faulty operations on the part of the person making the measurement. Acquiring proper work habits will tend to eliminate many mistakes. An *error* is a residual fault in the measuring instrument or in the technique of making the measurement. The use of a tape, the length of which is inaccurate, results in residual errors. Corrections may be made for such errors, and for extremely precise work many different corrections may need to be applied.

Cumulative and Compensating Errors

Cumulative errors are constant and uniform errors that affect measurements in the same manner; they consistently either increase or decrease the results of measurements and successively accumulate the error.

Compensating errors, to the contrary, are errors that tend to cancel each other. Whereas cumulative errors are either all plus or all minus, compensating errors in the same series of measurements are both plus and minus. Over a long distance, the ultimate result of cumulative errors in measurements may be considerable but, if the errors are compensating, the resulting error would be comparatively small.

Precision

Absolute accuracy of measurement is an ideal seldom, if ever, attained in reality. The degree of precision required depends on the character of work to be done. As an example, in laying out the wall lines for a building it may be considered sufficiently accurate if the measurements are correct to within $\frac{1}{8}$ in. In machining a delicate bearing, however, an error of $\frac{1}{1000}$ in. might be considered to be too great. Obviously, the more precise the measurement, the greater will be the cost of making it. The same care and precision used in surveying a plot of ground in the business center of a large city may be inappropriate for the survey of rural land of little worth.

As related to surveying, precision is the ratio of the error to the distance measured. This ratio is expressed as a fraction. If, for example, a precision of $\frac{1}{10\,000}$ is permitted, an error of 1 ft in a length of 10,000 ft would be acceptable. For farm and suburban surveying and for building layouts within the confines of a plot, a precision of $\frac{1}{5000}$ is usually considered to be acceptable. This is an error of approximately $\frac{1}{4}$ in. in 100 ft. City surveys generally require a precision of $\frac{1}{10\,000}$; sometimes greater precision is required. For this degree of precision the common builders' level is not sufficiently accurate and it becomes necessary to apply certain corrections and refinements, which are beyond the scope of this book.

Mistakes in Measuring Distances

Among the common mistakes made in measuring distances is the failure to observe the position of the zero mark on the tape. This is particularly likely to occur if the surveyor changes from one type of tape to another.

In measuring long distances a whole length of tape may be omitted. To avoid this, a pebble or coin is sometimes placed in a pocket for each full tape length recorded. Another aid is to have more than one individual keep the count.

In reading the tape the wrong foot mark may be read, such as 76.92 ft instead of 75.92 ft. Sometimes numbers are read upside down, 6 being mistaken for 9 or 68 instead of 89. Another very common mistake is to transpose figures, such as reading 21.51 instead of 21.15.

It is an excellent habit to estimate mentally the distance to be measured. Such a practice will tend to eliminate many of the larger discrepancies.

Errors in Measuring Distances

Failure to have the tape stretched tight while measuring results in a dimension that is too great. Permitting the tape to bend around bushes, boulders, etc., also results in a faulty measurement. Taking dimensions during a high wind permits the tape to be blown out of line, and this type of work should be avoided when weather conditions are unfavorable. Incorrect alignment of the tape, as explained in Section 4.5, is another common error, but this particular error may easily be guarded against. The above-mentioned errors all tend to make the recorded measurements too long; they are *cumulative*.

Failure to hold the plumb bob exactly over the point results in a dimension that is either too great or too small. Errors of this type are *compensating*, one error tending to correct a former error.

Tapes of Incorrect Length

One of the most serious sources of error lies in using a tape whose length is false. For example, the markings may indicate the length to be 100 ft, but, in reality, the true length of the tape may be somewhat more or less than this dimension. If such a tape is used, the resulting measurements will be in error, a *cumulative* error.

Surveyors frequently send one of the tapes in their office to the National Bureau of Standards to have its length checked. On the return of this tape, it is kept as a standard by means of which the other tapes are checked so that they may be used in the field.

If the error per tape length is known, a correction figure may be computed. This correction, added or subtracted from a measurement made with the faulty tape, will give the corrected dimension.

Rule 1. If the tape is longer than the standard, the correction must be added. Suppose that we have a tape marked to be 100 ft in length but whose true length is 110 ft. This, of course, is an exaggerated error and is used for the purpose of illustration; actually, the errors are usually fractions of an inch. Refer to Figure 4.6*a* and note that points *A* and *B* are exactly 100 ft apart. If we place the zero point of the faulty tape at point *A* and run out the entire length of the tape, the 100 ft mark on the tape will be beyond point *B*. The reading on the tape at point *B* would be at approximately 90.91 ft. Since this tape reading is less than the true dimension, a correction must be *added* to the tape reading to give the exact length.

Rule 2. If the tape is shorter than the standard, the correction must be subtracted. Let us assume now that we have a tape that, although marked 100 ft in length, has a true length of only 90 ft. In Figure 4.6*b* points *A* and *B* are exactly 100 ft apart. If the zero point on the inaccurate tape is placed at point *A* and the tape is extended its full length, the 100 ft mark on the tape will not reach point *B*. By use of the faulty tape the measurement from *A* to *B* would be 111.11 ft. Therefore, since this distance is too great, we must *subtract* a correction to the tape reading to find the true measurement.

Example. The distance between two points when measured with a 100 ft steel tape was found to be 323.52 ft. When compared with a standard, the length of the steel tape was found to be only 99.83 ft instead of 100 ft. Compute the true distance between the two points.

Solution:

$$100.00 \text{ ft} - 99.83 \text{ ft} = 0.17 \text{ ft}$$

hence,

correction = 0.17 ft per 100 linear ft

or,

correction = 0.0017 ft per linear ft

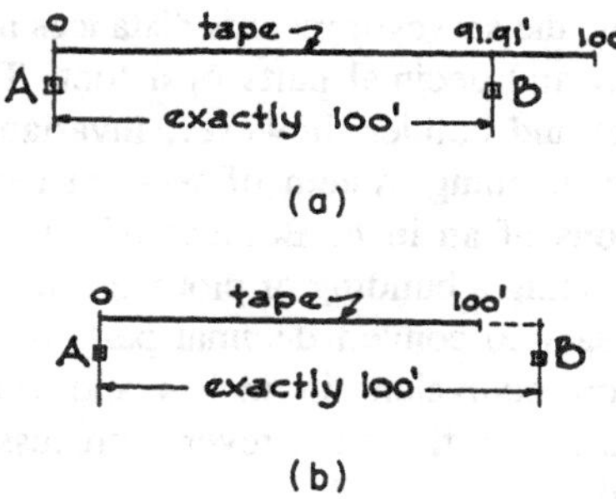

FIGURE 4.6.

The correction for the distance 323.52 ft is 323.52 × 0.0017, or 0.549984 ft, say 0.55 ft.

Because the tape is *too short*, the correction must be subtracted. Therefore,

$$323.52 - 0.55 = 322.97 \text{ ft} \quad \text{the true distance between the two points}$$

PROBLEM 4.7.A THROUGH F.

The following problems relate to measuring distances with inaccurate tapes. The second column gives the distances measured by the tape, and the third column gives the actual lengths of the tapes when compared with a 100 ft standard. For each condition compute the true distance.

Problem	Measured Distance	Actual Length of Tape
4.7.A	78.63 ft	100.13 ft
4.7.B	153.17 ft	100.13 ft
4.7.C	23.16 ft	100.09 ft
4.7.D	456.73 ft	99.71 ft
4.7.E	83.27 ft	100.33 ft
4.7.F	126.26 ft	99.94 ft

4.8 CONVERSIONS

Because surveyor's instruments are graduated in feet and decimal parts thereof, and because this system of graduations simplifies computations, the surveyor records distances in units of feet and decimal parts of a foot. The architect and builder, however, invariably use a dimensioning system of feet, inches, and fractions of an inch. Because of this, in dimensioning a building or plot plan, it is often necessary to convert decimal parts of a foot to their equivalent in inches and fractions thereof. Sometimes the reverse process is required.

For ordinary surveying, dimensions are given, or required, only to the nearest $1/100$ of a foot. For this degree of precision the following system of conversion will be found to be of great convenience. Certain equivalents are known as, for instance,

$3'' = 0.25'$ (exact)

$4'' = 0.33'$ (approximate)

$6'' = 0.50'$ (exact)

$8'' = 0.67'$ (approximate)

$9'' = 0.75'$ (exact)

Now, since $12'' = 1'$, $1'' = \frac{1}{12}'$, or $0.08333'$. Hence, $1'' =$ approximately $0.08'$ and, consequently, $\frac{1}{8}'' = 0.01'$, approximately. Using this equality, $\frac{1}{8}'' = 0.01'$, and the above equivalents, we may tabulate inches and their equivalents in decimals of a foot as follows:

$1'' = 0.08'$

$2'' = 0.17'$ (0.25 less 0.08)

$3'' = 0.25'$

$4'' = 0.33'$

$5'' = 0.42'$ (0.50 less 0.08)

$6'' = 0.50'$

$7'' = 0.58'$ (0.50 plus 0.08)

$8'' = 0.67'$

$9'' = 0.75'$

$10'' = 0.83'$ (0.75 plus 0.08)

$11'' = 0.92'$ (1.00 less 0.08)

$12'' = 1.00'$

As for fractions of an inch, we know that $\frac{1}{8}'' = 0.01'$. Then $\frac{1}{4}'' = (2 \times 0.01) = 0.02'$, $\frac{3}{8}'' = (3 \times 0.01) = 0.03'$, etc. Hence, for the decimal equivalent of inches and fractions of an inch, we add or subtract from the nearest whole inch in the above tabulation. For example,

$$3\tfrac{1}{8}'' = 0.25 + 0.01 = 0.26'$$

$$7\tfrac{3}{4}'' = 0.67 - 0.02 = 0.65'$$

$$5\tfrac{5}{8}'' = 0.50 - 0.03 = 0.47'$$

$$11\tfrac{1}{2}'' = 1.00 - 0.04 = 0.96'$$

Converting decimals of a foot to inches is done in a similar manner. Thus,

$$0.53' = 6 + \tfrac{3}{8} = 6\tfrac{3}{8}''$$

$$0.23' = 3 - \tfrac{1}{4} = 2\tfrac{3}{4}''$$

$$0.68' = 8 + \tfrac{1}{8} = 8\tfrac{1}{8}''$$

$$0.89' = 11 - \tfrac{3}{8} = 10\tfrac{5}{8}''$$

With practice these conversions may be made mentally and tables may be dispensed with.

Conversion Table

When distances are required to the nearest $\frac{1}{100}$ of a foot, or to the nearest $\frac{1}{8}''$, the method given above will give accurate results. If greater precision is desired, Table 4.1 may be used. This is a table in which inches and fractions of an inch are expressed in decimals of a foot, the equivalents being carried to the fourth decimal place. Note that the fractions advance by thirty-seconds of an inch. By referring to the table, we read directly that $4\tfrac{17}{32}'' = 0.3776'$, $8\tfrac{3}{16}'' = 0.6823'$, $0.6563' = 7\tfrac{7}{8}''$, $0.7552' = 9\tfrac{1}{16}''$, etc.

Metric Units

Work in this book is presented entirely in units of feet and inches (U.S. System). Conversions are also frequently necessary between these units and a metric-based system. While

TABLE 4.1 CONVERSION OF INCHES TO DECIMALS OF A FOOT

in.	0	1	2	3	4	5	6	7	8	9	10	11
0	ft	.0833	.1667	.2500	.3333	.4167	.5000	.5833	.6667	.7500	.8333	.9167
$\frac{1}{32}$	.0026	.0859	.1693	.2526	.3359	.4193	.5026	.5859	.6693	.7526	.8359	.9193
$\frac{1}{16}$	.0052	.0885	.1719	.2552	.3385	.4219	.5052	.5885	.6719	.7552	.8385	.9219
$\frac{3}{32}$	.0078	.0911	.1745	.2578	.3411	.4245	.5078	.5911	.6745	.7578	.8411	.9245
$\frac{1}{8}$	.0104	.0938	.1771	.2604	.3438	.4271	.5104	.5938	.6771	.7604	.8438	.9271
$\frac{5}{32}$	.0130	.0964	.1797	.2630	.3464	.4297	.5130	.5964	.6797	.7630	.8464	.9297
$\frac{3}{16}$	.0156	.0990	.1823	.2656	.3490	.4323	.5156	.5990	.6823	.7656	.8490	.9323
$\frac{7}{32}$	.0182	.1016	.1849	.2682	.3516	.4349	.5182	.6016	.6849	.7682	.8516	.9349
$\frac{1}{4}$	.0208	.1042	.1875	.2708	.3542	.4375	.5208	.6042	.6875	.7708	.8542	.9375
$\frac{9}{32}$	.0234	.1068	.1901	.2734	.3568	.4401	.5234	.6068	.6901	.7734	.8568	.9401
$\frac{5}{16}$	.0260	.1094	.1927	.2760	.3594	.4427	.5260	.6094	.6927	.7760	.8594	.9427
$\frac{11}{32}$	.0286	.1120	.1953	.2786	.3620	.4453	.5286	.6120	.6953	.7786	.8620	.9453
$\frac{3}{8}$	.0313	.1146	.1979	.2813	.3646	.4479	.5313	.6146	.6979	.7813	.8646	.9479
$\frac{13}{32}$	.0339	.1172	.2005	.2839	.3672	.4505	.5339	.6172	.7005	.7839	.8672	.9505
$\frac{7}{16}$	.0365	.1198	.2031	.2865	.3698	.4531	.5365	.6198	.7031	.7865	.8698	.9531
$\frac{15}{32}$	.0391	.1224	.2057	.2891	.3742	.4557	.5391	.6224	.7057	.7891	.8724	.9557
$\frac{1}{2}$	.0417	.1250	.2083	.2917	.3750	.4583	.5417	.6250	.7083	.7917	.8750	.9583
$\frac{17}{32}$	.0443	.1276	.2109	.2943	.3776	.4609	.5443	.6276	.7109	.7943	.8776	.9609
$\frac{9}{16}$	.0469	.1302	.2135	.2969	.3802	.4635	.5469	.6302	.7135	.7969	.8802	.9635
$\frac{19}{32}$	.0495	.1328	.2161	.2995	.3828	.4661	.5495	.6328	.7161	.7995	.8828	.9661
$\frac{5}{8}$	.0521	.1354	.2188	.3021	.3854	.4688	.5521	.6354	.7188	.8021	.8854	.9688
$\frac{21}{32}$	.0547	.1380	.2214	.3047	.3880	.4714	.5547	.6380	.7214	.8047	.8880	.9714
$\frac{11}{16}$	.0573	.1406	.2240	.3073	.3906	.4740	.5573	.6406	.7240	.8073	.8906	.9740
$\frac{23}{32}$	.0599	.1432	.2266	.3099	.3932	.4766	.5599	.6432	.7266	.8099	.8932	.9766
$\frac{3}{4}$	.0625	.1458	.2292	.3125	.3958	.4792	.5625	.6458	.7292	.8125	.8958	.9792
$\frac{25}{32}$	.0651	.1484	.2318	.3151	.3984	.4818	.5651	.6484	.7318	.8151	.8984	.9818
$\frac{13}{16}$	.0677	.1510	.2344	.3177	.4010	.4844	.5677	.6510	.7344	.8177	.9010	.9844
$\frac{27}{32}$	.0703	.1536	.2370	.3203	.4036	.4870	.5703	.6536	.7370	.8203	.9036	.9870
$\frac{7}{8}$	.0729	.1563	.2396	.3229	.4063	.4896	.5729	.6563	.7396	.8229	.9063	.9896
$\frac{29}{32}$	.0755	.1589	.2422	.3255	.4089	.4922	.5755	.6589	.7422	.8255	.9089	.9922
$\frac{15}{16}$	.0781	.1615	.2448	.3281	.4115	.4948	.5781	.6615	.7448	.8281	.9115	.9948
$\frac{31}{32}$	.0807	.1641	.2474	.3307	.4141	.4974	.5807	.6641	.7474	.8307	.9141	.9974
	0	1	2	3	4	5	6	7	8	9	10	11

the method of this conversion is not presented here, it is a common occurrence in present design work and is presented in many other references.

PROBLEMS 4.8.A AND B.

Perform the computations necessary for the following conversions and compare your results from that obtained from Table 4.1.

4.8.A. Convert the following linear dimensions to feet and decimals of a foot: 129 ft $2\frac{1}{2}$ in.; 75 ft $0\frac{5}{8}$ in.; 23 ft $9\frac{3}{4}$ in.; 351 ft $7\frac{5}{8}$ in.; 17 ft $4\frac{3}{8}$ in.; 183 ft $2\frac{1}{8}$ in.

4.8.B. Convert the following dimensions to feet and inches: 25.19 ft; 68.46 ft; 92.10 ft; 145.60 ft; 236.21 ft; 33.95 ft.

5

MEASURING ANGLES

Professional surveyors now use quite sophisticated equipment for the measurement of both horizontal and vertical angles. However, simple surveys and measurements are still done with the equipment described in this chapter, which is currently available from manufacturers. In any event, the basic surveying and measuring functions can be learned through the use of this simple equipment, which is most likely to be available for this purpose.

5.1 LEVELS AND TRANSITS

The builders' level and transit level are simplified versions of the more accurate instruments used by engineers or professional surveyors. Fundamentally, they are used for taking elevations and angles, and the basic functions of the builders' instruments are similar to those of the engineers' level and transit.

The Builders' Level

The essential parts of a *builders' level* are a *telescope* about 12 in. in length, a graduated *horizontal circle* divided into 1-degree spaces, an attached *vernier* to permit the reading of angles to 5 minutes, a *level vial*, and four *levelling screws* by means of which the instrument is brought to a level position. These basic parts are assembled into a single unit that is attached to a *tripod* with adjustable legs. Such an instrument is shown in Figure 5.1. A builders' level may be revolved in only a *horizontal plane*. It is the instrument most commonly used by persons other than professional surveyors.

The Builders' Transit

The builders' transit, (see Figure 5.2), contains all the features found in the builders' level but, in addition, the telescope is capable of being revolved in a vertical as well as in a

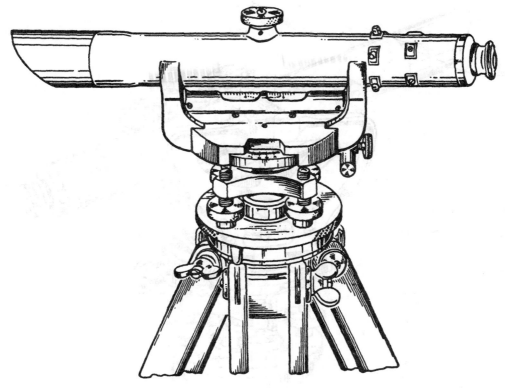

FIGURE 5.1.   The builder's level.

horizontal plane. To measure the vertical in-
clination a *vertical arc* is provided. This arc
is generally divided into 1-degree spaces, and
an accompanying vernier permits readings to
5 minutes. The transit permits the reading of
angles of both elevation and depression to
45°. When used on a sloping terrain, the
transit has an obvious advantage over the
level. Some builders' transits are equipped
with a compass. The surveying instruments
used by engineers and professional surveyors
permit greater accuracy and are more versa-
tile in their operation than the builders' levels
and transits.

5.2 BASIC COMPONENTS OF LEVELS AND TRANSITS

The following are some of the basic parts and
functions of levels and transits.

The Telescope

The telescope on levels and transits consists
of metal tubes in which the various lenses
common to the ordinary telescope are found.
The lens at the front end is the *objective*, and
the rear lens is contained in the *eye piece*. In
the telescope there is a ring on which spider-
web *cross hairs* are stretched at right angles
to each other. Focusing on the cross hairs is
accomplished by a spiral mechanism in the
eye piece. Focusing on the object to be sighted
is accomplished by a screw, on top of the instrument, that operates a lens located between
the objective and the cross hairs; this is known
as internal focusing.

The Spirit Level

Directly below the telescope, and parallel to
it, is the *bubble tube* in which a *spirit level* is

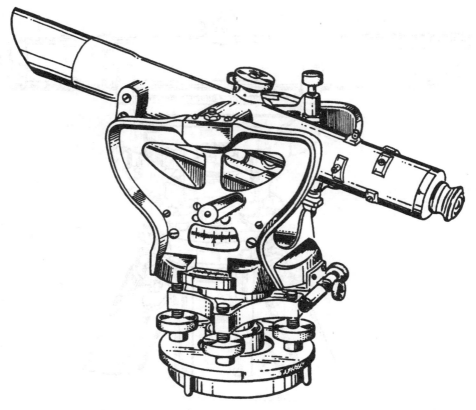

FIGURE 5.2.   The transit.

contained. See Figure 5.1. This consists of a glass tube sealed at both ends and almost completely filled with a nonfreezing fluid. The tube is either slightly bent or is a straight tube in which the upper inside surface is ground to a longitudinal circular curve. A scale is etched on the upper part of the glass tube, reading in both directions from the center. As the ends of the tube are raised or lowered the air bubble in the liquid takes various positions, and when the scale indicates that the bubble is exactly in the center of the tube, the tube, and consequently the telescope, are in a level position. For accuracy, the line of sight in the telescope and the spirit level must be parallel.

The Horizontal Circle

The horizontal circle on a builders' level or transit is generally graduated in intervals of

1° with each unit of 10° numbered continuously around the circle. Certain instruments have the numbering running both clockwise and counterclockwise. A thumb screw serves to clamp the telescope support to the graduated circle. By loosening the screw the telescope is permitted to revolve so that the object may be sighted. This screw is then tightened, and another thumb screw, called the *tangent screw*, is turned back and forth to permit the object to be sighted accurately.

Assume that the instrument is set up over a point, that the line of sight is on an object, and that the horizontal scale is set at 0°. By revolving the telescope and sighting on a second object the angle between the two objects may be found by reading the number of degrees on the graduated scale. Since the horizontal circle is graduated in degrees, the reading is only an approximation and a more accurate reading is obtained by the use of the vernier at the side of the horizontal circle.

The Vernier

A *vernier* is a short scale that is adjacent to the divisions of a graduated scale; its purpose is to determine the fractional part of the smallest units of the graduated scale. In Figures 5.3*a*, *b* and *c* a vernier is shown adjacent to a graduated scale. On the vernier is a zero point called the *index*; the vernier is an aid in reading the position of the index on the graduated scale.

Figure 5.3*a* shows a portion of a 1 in. scale with subdivisions of $\frac{1}{10}$ in. Below this scale is a vernier; it is $\frac{9}{10}$ in. length, and it also is divided into tenths. Consequently, each division of the vernier is $\frac{9}{10}$ of a division on the scale. In Figure 5.3*a* the index on the vernier coincides with the 0 mark on the scale, and hence the 1 mark on the vernier must be at a point $\frac{1}{10}$ of $\frac{1}{10}$ in. (or $\frac{1}{100}$ in.) to the left of the $\frac{1}{10}$ in. mark on the scale. The 2 mark on the vernier lies at $\frac{2}{100}$ in. to the left of the $\frac{2}{10}$ in. mark on the scale, and so on.

Now, let us move the vernier so that the 1 mark on the vernier coincides with the $\frac{1}{10}$ in. mark on the scale. This is shown in Figure 5.3*b*. Actually, we have moved the vernier $\frac{1}{100}$ in. to the right. If we had moved the vernier $\frac{2}{100}$ in. to the right, the 2 mark on the vernier would have coincided with the $\frac{2}{10}$ in. mark on the scale. The vernier permits us to divide the $\frac{1}{10}$ in. spaces on the scale into hundredths.

Suppose the vernier is moved to the position shown in Figure 5.3*c*, and we are asked to determine the exact position of the index on the vernier in relation to the adjacent 1 in. scale. We can see that it lies at some point

between 7.3 in. and 7.4 in. Now, since the 6 mark on the vernier coincides with one of the divisions on the scale, the index lies $\frac{6}{100}$ in. to the right of the 7.3 mark and the reading, therefore, is 7.36 in.

The rule for reading a vernier of this type is: *Record the nearest scale reading adjacent to the index on the vernier. Next, observe the number of the line on the vernier that coincides with one of the division marks on the scale, and add this number of the number previously recorded.*

Verniers Used in Measuring Angles

The principle of the vernier is applied to reading the divisions of a circular arc. The graduations of the horizontal circle and their accompanying verniers on surveying instruments and varied. Most builders' levels permit a reading to 5', whereas, with the more accurate surveyors' instruments, the readings may be made to 1' or even to 20". *The smallest angle that can be read with a vernier is called the least count*; it is equal to the smallest graduation on the circle divided by the number of divisions on the vernier.

Figure 5.4*a* shows a vernier and a portion of a horizontal circle that are found on some transits. The index on the vernier coincides with the 0 mark on the graduated circle. Note that 30 spaces on the vernier correspond to 29 spaces on the graduated circle. The smallest division on the circle is $\frac{1}{2}$ of 1°(30'), and the number of divisions on the vernier is 30; hence, $30' \times \frac{1}{30} = 1'$, the least count. The vernier shown in this figure extends both to the right and to the left of the index. Such a vernier is known as a *double-direct vernier*; it permits the reading of angles when the

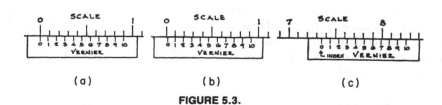

(a)   (b)   (c)

FIGURE 5.3.

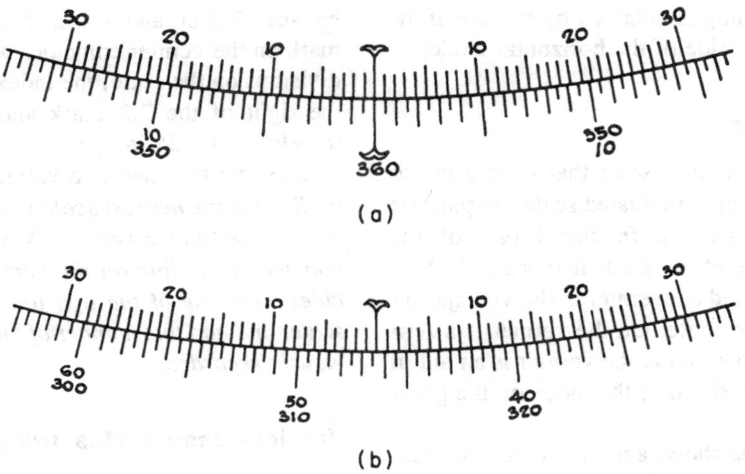

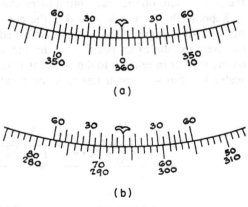

(a)

(b)

FIGURE 5.4.

telescope is revolved in either direction. The graduations on the circle are numbered both clockwise and counterclockwise. In revolving the telescope, the outer circle remains stationary while the inner circle (the vernier) revolves with the telescope.

Figure 5.4*b* shows the same circle and vernier when the telescope has been revolved to another position. It is important to remember that *the arrow, or index, on the vernier points to the angle that is to be read.* Since there are two verniers, one on each side of the index, the vernier to use in reading an angle is the one that extends beyond the index in *the same direction* in which the *increasing number* of degrees is read on the circle. Since two different angles may be read, the angle desired depends upon the direction in which the telescope has been revolved. Suppose, for instance, from the position shown in Figure 5.4*a*, the telescope had been revolved clockwise. Then the reading on the circle would be 46 degrees plus a certain number of minutes. Then, reading the vernier in the same direction (to the left of the index), we find that the line on the vernier that coincides with a line on the graduated circle is line 21. Therefore, the reading is 46 degrees plus 21 minutes, 46°21′.

Now suppose that the telescope had been revolved counterclockwise. The index points to 313 degrees 30 minutes plus some more minutes. Reading the vernier to the right of the index we see that the 9 mark on the vernier coincides with a graduation on the circle, and therefore the reading is 313°30′ plus 9″, or 313°39′. The sum of the two readings should equal 360°. Thus, 46°21′ + 313°39′ = 360°.

Figure 5.5*a* shows a circle and double vernier often found on a builders' transit. Note that the circle is divided into degrees (units of 60′) and that 12 spaces on the vernier equal 11 spaces on the graduated circle. This indicates that the least count is $60' \times \frac{1}{12}$, or 5′.

(a)

(b)

FIGURE 5.5.

Therefore, the smallest angle that can be read is to 5′.

Figure 5.5b shows the same circle and vernier when the telescope has been revolved to a new position. If the telescope has been turned clockwise, the index points to a reading of 66 degrees plus some minutes. Reading in the same direction (to the left of the index), we see that the fourth space on the vernier coincides with a space on the graduated circle. Therefore, since the least count is 5′, 4 × 5 = 20′; hence the reading is 66°20′. In a similar manner, if the telescope had been turned counterclockwise, the angle is read 293°40′.

The Compass

A part of an engineers' transit is the compass; it is also found on some builders' transits. The magnetic needle of the compass automatically points to the *north magnetic pole*. This is not the *true geographic pole*, and the angle between them is called the *magnetic declination*. The declination for a particular location may be found on government charts as, for example, 10°W, by which we mean that the compass points 10° west of the true or geographic north. All bearings given on surveys or site plans must be related to *the true north*; hence, whenever possible, the bearings on surveys should be related to some established line, the true bearing of which is known.

The pivot of the compass is at the center of a horizontal circle that is divided into quadrants, the quadrants being graduated into 90° each. To give a *bearing* of a certain point or line, it is necessary to give both the quadrant and the number of degrees. For example, a bearing of N 35°W indicates that the direction of an object sighted is on a line that is 35° west of north. See Section 6.2.

5.3 USING THE LEVEL AND THE TRANSIT

The following discussions treat some of the practical considerations for the use of the level and the transit.

Setting Up the Instrument

When the level or transit is to be set up over a given point, extreme care must be taken. Having attached the instrument to the tripod, the tripod is forced firmly into the ground so that the instrument is approximately over the point in the stake. The four levelling screws are operated so that the spirit level shows the instrument to be level. The plumb bob suspended from the instrument shows the relation of the instrument to the point in the stake, and a slight loosening of the levelling screws permits the head to be moved so that the plumb bob is brought directly over the point. When this has been accomplished the levelling screws are again operated until the spirit level indicates the instrument to be level in all directions.

Measuring Horizontal Angles

Suppose that the instrument has been set up over a given point and that we wish to measure the angle between two distant points, the point of the set-up being the vertex of the angle. The rodman goes to one of the points and holds a ranging pole in a vertical position directly over the point. The clamp screw on the instrument is loosened and the telescope is revolved so that the pole and cross hairs are in approximate alignment. The clamp screw is not tightened, and the tangent screw is operated to bring the pole and cross hairs into an exact line. The angle on the horizontal circle is now read and recorded.

The rodman now moves to the other point, the instrument is sighted on this point, and the angle is again read and recorded. The difference between the two readings is the angle between the two points. When a transit is used instead of a level, the instrument may be sighted directly on the points, eliminating the necessity of the ranging pole and ensuring greater accuracy. Many instruments permit the setting of 0 on the first point, thus obtaining the angle directly.

When the construction of the surveying instrument permits, angles may be measured with greater accuracy by using the *method of repetition*. Referring to Figure 5.6a, suppose we are required to measure the angle *BAC*. The procedure is as follows:

STEP 1.   Set up the instrument over point *A* with both clamps loose, and set 0 of the graduated circle and the index of the vernier approximately together. Tighten the *upper clamp*, and with the upper tangent screw bring the index and the 0 of the graduated circle together.

STEP 2.   With the upper clamp tight and the lower clamp loose, turn the instrument to sight approximately on point *B*. Tighten the *lower clamp*, and, with the lower tangent screw, set the vertical cross hair exactly on point *B*. The 0 of the instrument is now set on *B*.

STEP 3.   Loosen the *upper clamp*, and turn the telescope to sight on point *C*. Tighten the upper clamp, and, with the upper tangent screw, bring the vertical cross hair to coincide with point *C*. Record this reading.

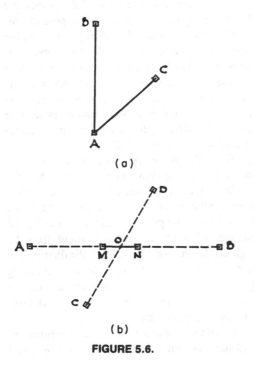

(a)

(b)

FIGURE 5.6.

STEP 4.   Now repeat Step 2. The angular value of *BAC* is now sighted on point *B*.

STEP 5.   Repeat Step 3. The reading now taken is twice the actual angle *BAC*. By dividing this reading by 2 we obtain the *average* of the two readings which, of course, is more accurate than the first reading taken in Step 3.

Setting Points on Line

When two points have been established, other points on the line between them may be readily determined. The instrument is set up and levelled over one of the given points sighted on the other point. The line of sight of the telescope is now in a vertical plane that passes through the two established points. To locate any other point on this line the rodman goes to the approximate position and moves the ranging pole to the right or left (as directed by the instrument man) until the image of the pole and the cross hairs are in alignment. The rodman drives a stake in the ground at this point. For accuracy, the operation is repeated and the rodman marks with a tack the point on the stake that is in exact alignment with the two given points.

Laying Off an Angle

Suppose that a line has been established and that it is required to lay off an angle of a given magnitude from some point on the line. First, the instrument is set up and levelled over the given point and the instrument is sighted on the established line. The angular reading on the horizontal circle is now recorded. To this angle we add (or subtract, as may be required) the given angle. The clamp screw on the instrument is loosened, and the telescope is turned until the index of the vernier points (approximately) to the sum (or difference) of the angles. Now the clamp screw is tightened, and the tangent screw is operated so that the proper line on the vernier coincides with a line on the graduated circle. The telescope has now been revolved the required angle, and a stake and tack are located on this line, thus establishing the required angle.

The method of laying off angles by repetition is explained in Section 13.1.

Intersections of Lines

A problem that frequency occurs in staking out buildings is to find the point of intersection of two previously established lines. Referring to Figure 5.6b, assume that points A, B, C, and D have been established and that we wish to find the point of intersection of lines AB and CD. First, set up the instrument over point A and sight on point B. The line of sight in the telescope now lies in a vertical plane that includes line AB. Stakes are now driven on line AB at points M and N. Points M and N are several feet apart and are located on the left and right sides of line CD, respectively. By means of the plumb bob, points M and N are accurately established in the stakes by tacks and a string is stretched between them. Now move to position C, set up the instrument, and sight on point D. This line of sight intersects the string at point O, at which point a stake is driven and, with the aid of a plumb bob, a tack is driven to indicate the intersection of lines AB and CD.

5.4 SUGGESTIONS

In a book of this scope detailed instructions for setting up and operating a level or transit are impracticable. The best method of acquiring facility in its use is to obtain an instrument, follow closely the accompanying instructions, and practice the various procedures. The following suggestions may prove to be helpful.

In setting up the instrument be certain that the legs of the tripod are firmly forced into the ground so that the instrument is not easily disturbed. Make certain that the cross hairs in the telescope are in sharp focus. Be sure that the plumb bob is exactly centered over the tack in the stake. While using the instrument, make frequent checks on the spirit level to see that the instrument has not been disturbed. When using a double vernier, be sure that the reading is made on the proper side of the index. If the horizontal circle has two rows of numbers in opposite directions, be careful to see that the reading is taken in the proper direction. In taking an angular reading of a circle, as illustrated in Figure 5.4b, do not forget to add the 30'; the reading is 313° + 30' + 9', 313°39', not 313°9'. Follow carefully the manufacturer's instructions relating to the care of the instrument.

6

SURVEYING METHODS
AND COMPUTATIONS

This chapter deals with the various problems of performing surveys and with the computations used in plotting and using them.

6.1 SURVEYS

Because of legal aspects, the original survey used for writing deed descriptions, or for staking out the boundary lines of a property, should be entrusted only to a registered surveyor. The architect or builder, before proceeding with any design or construction work, should require that the owner furnish this certified survey. Additional lines and grades may need to be determined within the boundary lines of the plot in order to locate buildings, roads, paths, etc. This work may be performed by the architect or builder if he is qualified. The architect does not make the original survey. However, in an illustrative example, consideration of which runs intermittently through this chapter, the complete computations are given to explain the proce-

dure and the various problems that are generally encountered. After the data concerning angles, length of lines, etc., have been obtained in the field, the necessary computations are made. Many methods may be employed in obtaining data, but the computations given in the illustrative example apply to all survey work.

Plane Surveying

Plane surveying treats the surface of the earth as a plane surface. Although this is not theoretically exact, the assumption is sufficiently accurate when surveys of comparatively small areas are involved.

6.2 ELEMENTS OF SURVEYS

Traverses

A *traverse* is a line or a series of connected lines surveyed across the earth's surface. An *open traverse* begins at a given point and ends at some distant point, as, for example, a survey of a highway or railroad. A *closed traverse* begins at a given point and returns to the same point, thus forming a closed circuit. This closed type of traverse is applicable to surveys of parcels of land, the boundaries forming a polygon.

Bearings of Lines

The bearing of a line is the horizontal angle between the direction of the line and a line pointing to the true north. For example, we say the bearing of a certain line is N 36° E. This indicates that the line is measured from the north in an easterly direction at an angle of 36°, as shown in Figure 6.1a. This bearing might also have been determined by measuring from the south, in which case the bearing would be recorded S 36° W. Bearings are always measured from either the north or south and from no other cardinal point, the angular value never being greater than 90°.

A practical method of relating the lines on a survey to the true north is to relate one of the lines to some line whose bearing has already been established. Such lines may be obtained from highway and city plans. If such a line forms a portion of the boundary of the plot, the bearings of the remaining lines are readily determined.

Example. In Figure 6.1b line *AB* has a bearing N 15° 25′ 20″ W. Line *BC* intersects *AB* at an angle of 42° 17′ 30″, as shown. Determine the bearing of the line *BC*.

Solution: For problems of this kind *always make a sketch* in which both lines are related to north, as shown in Figure 6.1c. By examining the sketch it is apparent that *BC* lies east of north and its bearing is the difference between 42° 17′ 30″ and 15° 25′ 20°, or N 26° 52′ 10″ E. The bearing of this line might also be given as S 26° 52′ 10″ W. The bearing is not always the difference between the two angles. It is necessary that the sketch be made, for the sketch indicates the proper procedure.

PROBLEMS 6.2.A THROUGH F.

In each of the diagrams shown in Fig. 6.2, two lines and the angles between them are shown, the bearing of one line being given. Compute the bearings of the remaining lines.

Intersecting Lines

Two nonparallel lines that intersect form four angles, as shown in Figure 6.3a. The opposite angles are equal; angle 1 = angle 3 and angle 2 = angle 4. Any two of the adjacent angles are *supplementary*, that is, their sum is equal to 180°.

We frequently have a problem in which the bearings of two lines are given, and we are required to find the angle between them. *Always* make a sketch, and note which angle is required. Do not confuse the required angle with its supplement. Is the required angle acute or obtuse, less or greater than 90°?

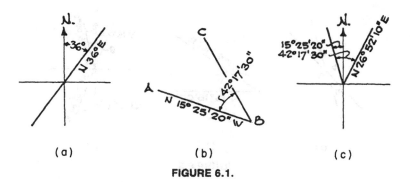

(a)   (b)   (c)

FIGURE 6.1.

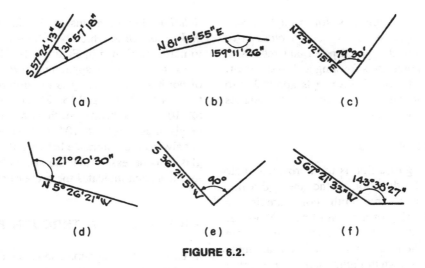

(a)   (b)   (c)

(d)   (e)   (f)

FIGURE 6.2.

Example. Two lines whose bearings are given are shown in Figure 6.3b. Determine θ, the angle of intersection.

Solution: A sketch is made, showing the bearings of the lines with relation to the points of the compass, Figure 6.3c. From Figure 6.3b we see that θ, the required angle, is the acute angle. Therefore, in Figure 6.3c, we extend the line whose bearing is S 21° 21′ 10″ E and we see that to obtain the acute angle we must add the given angles. Hence,

21° 21′ 10″

plus 13° 15′ 00″

34° 36′ 10° = θ the required angle

Example. Two intersecting lines whose bearings are given are shown in Figure 6.4a. Determine angle θ.

Solution: Figure 6.4b is the sketch showing the positions of the lines and their angles in relation to the points of the compass. From Figure 6.4a note that θ, the required angle, is the obtuse angle. Then

180° 00′ 00″

plus   21° 32′ 36″

201° 32′ 36″

minus   83° 23′ 02″

118° 0.9′ 34″ = θ the required angle

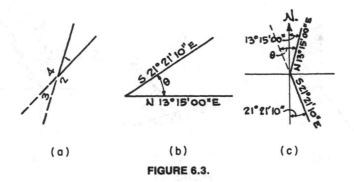

(a)   (b)   (c)

FIGURE 6.3.

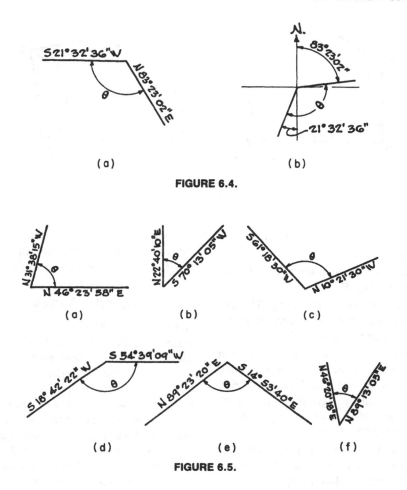

FIGURE 6.4.

FIGURE 6.5.

PROBLEMS 6.2.G THROUGH L.

In each of the diagrams shown in Figure 6.5 two intersecting lines and their bearings are given. For each pair of lines, compute the required angle in degrees, minutes, and seconds.

6.3 MAKING A SURVEY

To make a survey of an existing plot of ground the corners must first be definitely located and marked. If possible, the instrument should be set up directly over these points. In the following illustrative example it is assumed that this can be done. The method of making the survey described here is but one of several; it is particularly applicable when the builders' level or transit is used. Let us assume that the plot to be surveyed has the corners located as shown in Figure 6.6a. In moving the instrument from point to point we will take the points in this sequence: A, B, C, D, E. The angles measured will be the interior angles, and the instrument will always be turned to the right (clockwise direction) in reading the angles. The bearing of line AB, N 75° 20′ E, has been obtained from the township surveyor. To make the survey, the following steps are taken.

STEP 1. Set up the instrument over any point, say point A, and sight on point E.

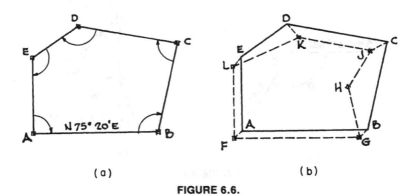

(a)   (b)

FIGURE 6.6.

STEP 2. Record the angular reading on the horizontal graduated circle.

STEP 3. Revolve the instrument (clockwise) and sight on point B.

STEP 4. Take the angular reading. The difference between the two recorded angles will be the internal angle at A.

STEP 5. Measure distance AE. For greater accuracy the distance should be measured twice, first from A to E and then from E to A. If there is only a slight difference in the measurements, their average may be taken as the length of the line

STEP 6. The instrument is now moved to point B, sights are taken on points A and C, and the above procedure is repeated.

This process is repeated at each successive point so that all the interior angles and the lengths of lines have been measured. For greater accuracy the entire procedure may be repeated and the measurements of the angles averaged.

Figure 6.7 shows a suggested form for recording the notes of the survey. Field books with ruled pages may be obtained for this purpose. In some books the left-hand page is used for the data shown in Figure 6.7 and the right-hand page is used for sketches and notes.

The survey shown in Figure 6.7 was made with an instrument having the horizontal circle graduated and marked from 0° to 360°. Some instruments are graduated in 90° quadrants and are so constructed that the horizon-

tal circle may be set at 0, thereby eliminating the first reading and the subsequent subtraction necessary to obtain the angle.

Checking Interior Angles

Before leaving the site, it is recommended to make an additional check on the accuracy of the angles. From plane geometry we know that in any closed polygon the sum of the interior angles equals $180° \times (n - 2)$, where n is the number of sides of the polygon. In this example the polygon has five sides and therefore the sum of the interior angles should equal $180 \times (5 - 2) = 180 \times 3$, or 540°.

Note in Figure 6.7 that the sum of the angles does not equal 540°; it is 539° 50′, a difference of 10′. This is usually the case. When using the builders' level or transit, which has an accuracy of only 05′ in reading angles, the discrepancy should not exceed $05' \sqrt{n}$. In this instance $n = 5$; hence $05' \times \sqrt{5} = 11.18'$. Since the difference of 10′ falls within the allowable limit, this indicates acceptable work considering the accuracy of the instrument.

Balancing Interior Angles

The discrepancy of 10′, explained in Section 6.3, must be distributed among the five angles. It might appear logical to distribute the 10′ equally among each of the angles, adding 02′ to each. This would result in angles such as 86° 42′, implying that the survey had been

ON STATION	SIGHT AT	ANGULAR READING	INTERNAL ANGLE	DISTANCE	AVERAGE DISTANCE
A	E	177° 15'	90° 00'	62.58	62.60'
	B	267° 15'		101.70	
B	A	126° 20'	102 ~~25~~ 30'	102.10	101.90
	C	228° 45'		75.03	
C	B	321° 15'	86° 40'	74.91	74.97
	D	47° 55'		84.41	
D	C	72° 20'	136° ~~15~~ 20'	84.55	84.48
	E	208° 35'		42.70	
E	D	102° 00'	124° 30'	42.68	42.69
	A	226° 30'		62.62	

Sum of Internal Angles = 539° 50'

N = __5__   180(N − 2) = 540°

FIGURE 6.7. Recorded data for a survey.

made with an instrument measuring to 0.1'. This, however, is known to be untrue; hence we will make the corrections by adding 05' to each of the angles at *B* and *D*, as shown in Figure 6.7. These two angles have been chosen arbitrarily. An assumption often made is that the angles with the shortest sides will have the greatest errors. If conditions in the field indicate that an error may have been made at some other angle or angles, the correction should be made at these points.

Alternative Positions of the Instrument

When the condition arises where obstructions prevent placing the instrument directly over the corners of the plot, stakes are driven at adjacent points, *F*, *G*, *H*, *J*, *K*, and *L*, as indicated in Figure 6.6*b*. These points are selected to permit an unobstructed view of each other as well as of the corners of the plot. A closed traverse is made of these points, and lines *FA*, *GB*, *JC*, etc., are run to the corner points. These lines and the angles they make with the traverse are measured. With this information the coordinates of points *A*, *B*, *C*, *D*, and *E* may be determined (see Section 6.4) and the boundaries of the plot thus established.

6.4 COMPUTATIONS FOR SURVEYS

Chapter 5 explained the procedure of determining the angles and the lengths of lines in surveying a plot of ground. These data are obtained at the site. The next step is to perform certain computations that enable one to plot the survey and to determine its area. The computations are generally made in the office.

Plotting the Survey

From the data found in the field, shown in Figure 6.7, the plot is drawn to scale, Figure 6.8. This diagram shows the internal angles, and the lengths and the bearings of all lines. The bearings are computed as explained in Section 6.2. From the township surveyor we found that the bearing of *AB* is N 75° 20' E. The angle between *AB* and *BC* is 102° 30'; hence the bearing of *BC* is S 2° 10' E. The bearings of the other sides are likewise determined. A protractor may be used in laying off the angles, but more accurate methods of plotting are explained in Section 6.5. Note in Figure 6.8 that the lines have been drawn in relation to true north as a vertical axis. This should always be done since it simplifies computations.

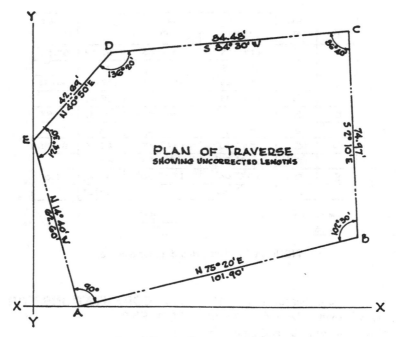

FIGURE 6.8. Plan of the traverse.

Latitudes and Departures

The *latitude* of a line is its projection on the north and south line. The *departure* of a line is its projection on the east and west line. They are the vertical and horizontal coordinates, as shown in Figure 6.9. In this figure *AB is the given line and its bearing is the angle θ.* Note that the angle at *B* is also angle θ. The length of line *AB* is 97.23′. In the right triangle observe the following relationships:

$$\text{departure} = \text{length} \times \sin \text{bearing}$$

$$\text{latitude} = \text{length} \times \cos \text{bearing}$$

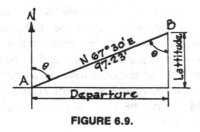

FIGURE 6.9.

By making sketches, it is seen that these relationships are valid regardless of the quadrant in which the line is drawn.

For line *AB* shown in Figure 6.8,

$$\text{departure} = 97.23 \times \sin 67° \, 30' = 89.83'$$

$$\text{latitude} = 97.23 \times \cos 67° \, 30' = 37.21'$$

The latitude and departure of all the sides of the plot shown in Figure 6.8 are computed as shown in Table 6.1. After they are computed, the latitudes and departures can be checked approximately by scaling the corresponding distances in Figure 6.8. This check will disclose any large errors in computations.

In traversing a route on a plane surface and returning to the starting point, a closed traverse, we must travel the same distance north as we do south and just as far east as west. Let us consider the distances travelled north and south as plus latitudes and minus latitudes, respectively, and distances travelled east and west as plus departures and minus

TABLE 6.1 LATITUDES AND DEPARTURES FOR THE EXAMPLE

Line	Length L (ft)	Bearing θ	Latitude $L \cos \theta$ (ft)	Departure $L \sin \theta$ (ft)
AB	101.90	N 75° 20′ E	25.80	98.58
BC	74.97	S 2° 10′ E	74.92	2.834
CD	84.48	S 84° 30′ W	8.097	84.09
DE	42.69	N 40° 50′ E	32.30	27.91
EA	62.60	N 14° 40′ W	60.56	15.85

departures. Then, *in any closed polygon or traverse, the plus latitudes must equal the minus latitudes and the plus departures must equal the minus departures.*

The latitudes and departures computed in Table 6.1 are tabulated in Figure 6.10 with respect to their plus and minus signs. Note in Figure 6.8 that in going from *A* to *B* we travel northward and eastward; therefore, both latitude and departure are placed in the plus columns. From *B* to *C* we go northward and westward, a plus latitude and minus departure. The other values are plotted similarly. We find, in Figure 6.10, that the sum of the plus latitudes does not equal the sum of the

minus latitudes. This is true also with respect to the departures; it indicates that the polygon does not close. It is improbable that these plus and minus values will balance in any survey. However, in order to have a polygon that closes mathematically, the plus and minus columns must be balanced and the lengths of the lines correspondingly adjusted.

Error of Closure

If we had accurately plotted the traverse shown in Figure 6.8 to a large scale, beginning at *A* and continuing in a counterclockwise direction, we would find that the end of

LINE	LATITUDES +	LATITUDES −	DEPARTURES +	DEPARTURES −	BALANCED LATITUDES	BALANCED DEPARTURES
AB	25.80		98.58		25.83	98.75
BC	74.92			2.83	75.01	2.83
CD		8.10		84.09	8.09	83.94
DE		32.30		27.91	32.26	27.86
EA		60.56	15.85		60.49	15.88
	100.72	100.96	114.43	114.83		
		100.72		114.43		
		.24		.40		

Error of Closure = $\sqrt{(.24)^2 + (.40)^2}$ = .47

Precision = $\dfrac{.47}{366.64}$ = $\dfrac{1}{780}$

CORRECTIONS

	LAT.	DEP.
AB	.03	.17
BC	.09	.00
CD	.01	.15
DE	.04	.05
EA	.07	.03
	.24	.40

FIGURE 6.10. Coordinates of the corner points.

the last line would not coincide with point A from which we started. We would have had the condition shown in Figure 6.11. The distance between the points A' and A, the length of line by which the polygon fails to close, is called the *error of closure*. Since the differences in the plus and minus latitudes and departures are shown in Figure 6.10 to be 0.24 and 0.40, respectively, the error of closure is the hypothenuse ($A'A$) of the right triangle, of which 0.24 and 0.40 are the other two sides as shown in Figure 6.11. Then,

$$\text{hypothenuse} = \sqrt{0.24^2 + 0.40^2}$$
$$= \sqrt{0.2176}$$
$$= 0.47' \quad \text{the error of closure}$$

Precision

Some error of closure is to be expected but, assuming there is no mistake in the computations, a large error indicates that a mistake has occurred in making the survey. The degree of error varies with the length of the lines and the accuracy of the instrument that is used.

The *precision of the survey* is equal to the error of closure divided by the sum of the lengths of the lines as expressed by a fraction with unity in the numerator. In our survey,

$$\text{precision} = \frac{0.47}{366.64} = \frac{1}{780}$$

The precision to be expected in a survey made with a builders' level or transit, measuring to the nearest 05', is about 1/500. In

our survey the precision is somewhat greater, indicating that the work was performed acceptably. A precision of 1/500 is suitable for farm surveying, and it is sufficiently accurate for laying out buildings, roads, etc., within the plot. In city surveying greater precision is required and precisions of 1/10,000 and 1/20,000 are often obtained.

Corrections to Latitudes and Departures

Since the latitudes and departures must be balanced, corrections must be made. The corrections in latitude and departure are distributed proportionately among *all sides* of the survey. The correction to be applied to the latitude of each side will be

$$\frac{\text{total error in latitude} \times \text{latitude of the side}}{\text{sum of the latitudes of all sides}}$$

Therefore, the correction to be applied to the latitude of line AB will be

$$\frac{0.24 \times 25.80}{201.68} = 0.0307 \quad \text{say } 0.03'$$

The correction need only to be taken to the nearest 1/100 of a foot.

The corrections to the departures are found in a similar manner. The corrections to be applied to the departure of each side will be

$$\frac{\text{total error in departure} \times \text{departure of the side}}{\text{sum of the departures of all sides}}$$

Hence, the correction to be applied to the departure of line AB will be

$$\frac{0.40 \times 98.58}{229.26} = 0.173 \quad \text{say } 0.17'$$

The corrections to latitudes and departures for all the sides of the survey are computed similarly.

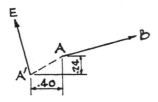

FIGURE 6.11.

Balancing Latitudes and Departures

The magnitudes of the corrections having been computed, they must be applied to the latitudes and departures shown in Figure 6.10. In applying the corrections, *subtract the corrections from the values in the columns having the greater sums* and *add the corrections to the values in the columns having the smaller sums.* This procedure has been followed and the corrected (balanced) values are shown in the last two columns of Figure 6.10. By checking, it is seen that the plus and minus latitudes and departures now balance.

Corrected Lengths and Bearings

The changes in the magnitudes of the latitudes and departures necessitate slight changes in the lengths of the sides of the survey and, possibly, changes in their bearings. These adjusted lengths are now computed by use of these two equations:

$$\text{length} = \sqrt{\text{latitude}^2 + \text{departure}^2}$$

$$\tan \text{bearing} = \frac{\text{departure}}{\text{latitude}}$$

To verify these relationships, see Figure 6.9.

The revised lengths are shown in the following computations, but the complete logarithmic work is shown only for line *AB*.

Line *AB*

$$\text{length} = \sqrt{\text{latitude}^2 + \text{departure}^2}$$
$$\text{length} = \sqrt{25.83^2 + 98.75^2}$$
$$= 102.07 \text{ ft}$$

$$\tan \text{bearing} = \frac{\text{departure}}{\text{latitude}} = \frac{98.75}{25.83}$$

$$\text{bearing} = \text{N } 75° \ 20' \ \text{E}$$

Line *BC*

$$\text{length} = \sqrt{75.01^2 + 2.83^2}$$
$$= 75.07 \text{ ft}$$

$$\tan \text{bearing} = \frac{2.83}{75.01}$$

$$\text{bearing} = \text{S } 2° \ 10' \ \text{E}$$

Line *CD*

$$\text{length} = \sqrt{8.09^2 + 83.94^2}$$
$$= 84.33 \text{ ft}$$

$$\tan \text{bearing} = \frac{83.94}{8.09}$$

$$\text{bearing} = \text{S } 84° \ 30' \ \text{W}$$

Line *DE*

$$\text{length} = \sqrt{32.26^2 + 27.86^2}$$
$$= 42.63 \text{ ft}$$

$$\tan \text{bearing} = \frac{27.86}{32.26}$$

$$\text{bearing} = \text{N } 40° \ 50' \ \text{W}$$

Line *EA*

$$\text{length} = \sqrt{60.49^2 + 15.88^2}$$
$$= 62.54 \text{ ft}$$

$$\tan \text{bearing} = \frac{15.88}{60.49}$$

$$\text{bearing} = \text{N } 14° \ 40' \ \text{W}$$

The corrected lengths of the sides are shown in Figure 6.18. In the above computations note that the bearings did not change by an amount sufficient to affect the bearing to the nearest 05'. This will generally be the case when the error of closure is within the limit previously noted. In these computations we have gone around the traverse in a counterclockwise direction because the original survey notes were designated in this manner. A clockwise direction might have been taken; the results would have been the same.

Coordinates of the Survey

If any other lines or points within the plot are to be determined, or if the area of the plot is to be computed, it is necessary to know the X and Y coordinates of all the corners of the surveys.

Referring to Figure 6.12, the X coordinate of point A is its distance from the Y-Y axis; this is identified as distance X_A, and it is called the *abscissa*. Similarly, the Y coordinate of point A is Y_A, the *ordinate*.

The X and Y coordinates of the corners of the survey are shown in Table 6.2. On referring to Figure 6.8, it is seen that the X coordinate of point A is equal to the departure of line EA; the X coordinate of point B is the departure of line EA plus the departure of line AB, and so on. It should be noted that the X-X and Y-Y axes for this survey have been arbitrarily chosen to pass through the most westerly and most southerly points of the perimeter of the survey.

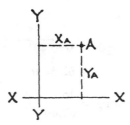

FIGURE 6.12.

TABLE 6.2 X AND Y COORDINATES FOR THE EXAMPLE

Point	X Coordinate	Y Coordinate
A	15.88	0
	+98.75	25.83
B	114.63	25.83
	−2.83	+75.01
C	111.80	100.84
	−83.94	−8.09
D	27.86	92.75
	−27.86	−32.26
E	0	60.49

6.5 PLOTTING THE SURVEY

After the lengths of the boundaries of the plot have been determined, as well as the angles between the intersecting sides, these data must be plotted on the drawing board. To accomplish this several methods are available.

(a) *The Protractor.* To employ this method the angles between the intersecting sides of the boundary are laid off in accordance with the divisions, indicating the degrees, marked on the protractor. The lengths of the sides are then laid off to scale on the appropriate sides. If great precision is required it cannot be obtained by this procedure. The *adjustable triangle* is sometimes used for laying off angles, but it too does not afford a great degree of accuracy. The use of a *vernier protractor* permits greater accuracy than that obtained with the ordinary protractor.

(b) *Plotting Coordinates.* When the X and Y coordinates of the corners of the survey have been computed, they may be accurately plotted to the desired scale and the lines of the survey are thus established. For this work make certain that the triangle used has a *true right angle*.

(c) *Plotting by Tangents.* The method of plotting by tangents affords a great degree of accuracy and is simple in its application. It requires a scale divided into inches and tenths of an inch (an engineer's scale) and a table of *natural tangents*. Table 6.3 is a table giving the natural tangents of angles up to and including 45°.

The examples that follow are solved using values from Table 6.3. However, the work can more easily be performed with a calculator capable of performing trigonometric operations. Angles must be entered in decimal form in such work (23.333°, not 23° 20′), so a conversion must be made to correspond to the values in Table 6.3.

TABLE 6.3 NATURAL TANGENTS

Angle	0'	10' 0.1667°	20' 0.3333°	30' 0.5°	40' 0.6667°	50' 0.8333°	60'
0°	0.00000	0.00291	0.00582	0.00873	0.01164	0.01455	0.01746
1°	0.01746	0.02036	0.02328	0.02619	0.02910	0.03201	0.03492
2°	0.03492	0.03783	0.04075	0.04366	0.04658	0.04949	0.05241
3°	0.05241	0.05533	0.05824	0.06116	0.06408	0.06700	0.06993
4°	0.06993	0.07285	0.07578	0.07870	0.08163	0.08456	0.08749
5°	0.08749	0.09042	0.09335	0.09629	0.09923	0.10216	0.10510
6°	0.10510	0.10805	0.11099	0.11394	0.11688	0.11983	0.12278
7°	0.12278	0.12574	0.12869	0.13165	0.13461	0.13758	0.14054
8°	0.14054	0.14351	0.14648	0.14945	0.15243	0.15540	0.15838
9°	0.15838	0.16137	0.16435	0.16734	0.17033	0.17333	0.17633
10°	0.17633	0.17933	0.18233	0.18534	0.18835	0.19136	0.19438
11°	0.19438	0.19740	0.20042	0.20345	0.20648	0.20952	0.21256
12°	0.21256	0.21560	0.21864	0.22169	0.22475	0.22781	0.23087
13°	0.23087	0.23393	0.23700	0.24008	0.24316	0.24624	0.24933
14°	0.24933	0.25242	0.25552	0.25862	0.26172	0.26483	0.26795
15°	0.26795	0.27107	0.27419	0.27732	0.28046	0.28360	0.28675
16°	0.28675	0.28990	0.29305	0.29621	0.29938	0.30255	0.30573
17°	0.30573	0.30891	0.31210	0.31530	0.31850	0.32171	0.32492
18°	0.32492	0.32814	0.33136	0.33460	0.33783	0.34108	0.34433
19°	0.34433	0.34758	0.35085	0.35421	0.35740	0.36068	0.36397
20°	0.36397	0.36727	0.37057	0.37388	0.37720	0.38053	0.38386
21°	0.38386	0.38721	0.39055	0.39391	0.39727	0.40065	0.40403
22°	0.40403	0.40741	0.41081	0.41421	0.41763	0.42105	0.42447
23°	0.42447	0.42791	0.43136	0.43481	0.43828	0.44175	0.44523
24°	0.44523	0.44872	0.45222	0.45573	0.45924	0.46277	0.46631
25°	0.46631	0.46985	0.47341	0.47698	0.48055	0.48414	0.48773
26°	0.48773	0.49134	0.49495	0.49858	0.50222	0.50587	0.50953
27°	0.50953	0.51320	0.51688	0.52057	0.52427	0.52798	0.53171
28°	0.53171	0.53545	0.53920	0.54296	0.54673	0.55051	0.55431
29°	0.55431	0.55812	0.56194	0.56577	0.56962	0.57348	0.57735
30°	0.57735	0.58124	0.58513	0.58905	0.59297	0.59691	0.60086
31°	0.60086	0.60483	0.60881	0.61280	0.61681	0.62083	0.62487
32°	0.62487	0.62892	0.63299	0.63707	0.64117	0.64528	0.64941
33°	0.64941	0.65355	0.65771	0.66189	0.66608	0.67028	0.67451
34°	0.67451	0.67875	0.68301	0.68728	0.69157	0.69588	0.70021
35°	0.70021	0.70455	0.70891	0.71329	0.71769	0.72211	0.72654
36°	0.72654	0.73100	0.73547	0.73996	0.74447	0.74900	0.75355
37°	0.75355	0.75812	0.76272	0.76733	0.77196	0.77661	0.78129
38°	0.78129	0.78598	0.79070	0.79544	0.80020	0.80498	0.80978
39°	0.80978	0.81461	0.81946	0.82434	0.82923	0.83415	0.83910
40°	0.83910	0.84407	0.84906	0.85408	0.85912	0.86419	0.86929
41°	0.86929	0.87441	0.87955	0.88473	0.88992	0.89515	0.90040
42°	0.90040	0.90569	0.91099	0.91633	0.92170	0.92709	0.93252
43°	0.93252	0.93797	0.94345	0.94896	0.95451	0.96008	0.96569
44°	0.96569	0.97133	0.97700	0.98270	0.98843	0.99420	1.00000

Example 1. Line *AB* in Figure 6.13*a* is horizontal. Let it be required to lay off an angle of 23° 20′, the angle *BAD*.

Solution: On referring to Table 6.3 we find that the natural tangent of 23° 20′ is 0.431. Now lay off *AC*, a length of 10 in. At point *C* erect a vertical line and on this line measure a distance 10 × 0.431 or 4.31 in.; call it point *D*. Then, since tan angle *DAC* = *DC*/*AC*, or 4.31/10, the line from *A* to *D* makes an angle of 23° 20′ with line *AB*. For greater accuracy, *AC* might have been made 20 in., in which case *CD* would be 20 × 0.43136, or 8.63 in.

Example 2. In Figure 6.13*b* the line *AB* is horizontal. Let it be required to lay off an angle of 83° 35′, the angle *BAD*.

Solution: From a table of natural tangents we find the tangent of 83° 35′ to be 8.8934. By the procedure used in the previous example, the vertical line would be 10 × 8.8934, or 88.934 in. Obviously, this length is too great to measure on the drawing board. Therefore, *when the angle to be laid off exceeds 45°, lay off the complement of the angle.* The complement of 83° 35′ = 90° 0′ − 83° 35′, or 6° 25′. Table 6.3 shows the natural tangent of

6° 25′ to be approximately 0.112. Now, in Figure 6.13*b*, lay off *AC*, 10 in. in length, perpendicular to *AB*, and from point *C* lay off *CD*, a horizontal line 1.12 in. in length. Thus, since angle *CAD* is 6° 25′, angle *BAD* will be 83° 35′, as required.

Having plotted the corners of the plot by the method of tangents, these points may be checked by measuring the coordinates of the points. This procedure will disclose any great errors.

6.6 DEED DESCRIPTIONS

Frequently it is necessary to plot a parcel of ground from its description in a deed. Deeds of all properties may be found in the office of the Recorder of Deeds located in the county seat or in the municipal offices. In each deed is a description of the property involved. The following deed description relates to the illustrative example for which the computations have been previously given in this book and which is shown in Figure 6.18.

All That Certain lot or piece of ground with the buildings and improvements thereon erected. *Situate* in Newville, Bucks County, Pennsylvania, beginning at the northeast intersection of Main Street (eighty feet wide) and Chestnut Street (fifty feet wide) and running along the easterly side of said Chestnut Street and measured N 14° 40′ W a distance of sixty-two and fifty-four hundredths (62.54) feet to a point at the southwest corner of property now or formerly belonging to William A. Weaver; thence turning and running by the land of said Weaver and measured N 40° 50′ E a distance of forty-two and sixty-three hundredths (42.63) feet to a point in the southern boundary of the said Weaver property; thence turning and running by the land of said Weaver and measure S 84° 30′ W a distance of eighty-four and thirty-three hundredths (84.33) feet to a point in the western boundary of a property now or formerly belonging to Robert B. Rogers; thence turning and running by the land of said Rogers and measured S 2° 10′ E a distance of seventy-five and seven hundredths (75.07) feet to a point in the northerly side of the aforesaid Main Street; thence turning and running along the northerly side of said Main Street and

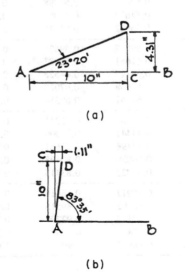

(a)

(b)

FIGURE 6.13.

measured N 75° 20′ E a distance of one hundred two and seven hundredths (102.07) feet to the first mentioned point and place of beginning.

6.7 COMPUTATION OF AREAS

It is frequently necessary to compute the area of a plot. There are several methods by which this may be accomplished; among them is the *method of coordinates*. In order to apply this method of finding the area of a closed traverse it is necessary that the coordinates of the corners of the plot be established. This procedure is explained in Section 6.4, and the X and Y coordinates for the corners of our illustrative example are tabulated in Table 6.3.

Figure 6.14 shows a four-sided area, *ABCD*, with the X and Y coordinates of points A and D. The coordinates of points B and C are not indicated. By examining this diagram it is seen that the area of *ABCD* may be computed by finding the area of trapezoid $DD'C'C$ and subtracting from it the sum of trapezoids $DD'A'A$, $AA'B'B$, and $BB'C'C$. The area of a trapezoid is equal to one-half the sum of the parallel sides multiplied by the perpendicular distance between them. By applying this system of computation, an equation for determining the area of the quadrilateral *ABCD* may be derived. A convenient form of the equation is:

$$\text{area of } ABCD = \tfrac{1}{2}[X_A(Y_B - Y_D) + X_B(Y_C$$

$$- Y_A) + X_C(Y_D - Y_B) + X_D(Y_A - Y_C)]$$

Therefore, to find the area of a polygon, begin at any corner (point A, for example) and proceed in a counterclockwise direction (A, B, C, D). Multiply the X coordinate of the beginning point (X_A) by the difference between the following and the preceding Y coordinates ($Y_B - Y_D$). Similarly, obtain the sum of these products for each point in succession. One-half this sum gives the area of the given polygon. The above equation relates to a four-sided figure, but this method of computation may be applied to a polygon having any number of sides.

The plot of ground in our example, shown in Figure 6.18, has five corners (five sides); hence, in the above equation, we must include the coordinates of points A, B, C, D, and E. These values are shown in Table 6.3 and are again tabulated in Figure 6.15. The above equation is most readily solved by tabulating the various terms as shown. The plus and minus areas, shown in the last two columns on the right side of the table, are the products of the Y differences and the X coordinates, as required by the formula. For example, the minus 550 shown in the last column of the table is $X_A(Y_B - Y_E)$, the product is $15.88 \times (-34.66)$. The tabulation shows that the area of the polygon *ABCDE* is 8,683 ft^2, or 0.1993 acres.

Another example of computing an area by this method is given in Section 8.5.

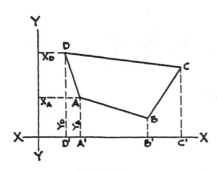

FIGURE 6.14.

COMPUTATION OF AREA

$$\text{Area} = \tfrac{1}{2}\Big[X_A(Y_B - Y_E) + X_B(Y_C - Y_A) + X_C(Y_D - Y_B) + X_D(Y_E - Y_C) + X_E(Y_A - Y_D)\Big]$$

COR.	COORDINATES		Y-DIFFERENCE		AREAS	
	X	Y	CORNERS	VALUE	+	−
A	15.88	0	B−E	−34.66		550
B	114.63	25.83	C−A	+100.84	11559	
C	111.80	100.84	D−B	+66.92	7482	
D	27.86	92.75	E−C	−40.35		1124
E	0	60.49	A−D	−92.75	0	

$$+19041 \quad -1674$$
$$-1674$$
$$\overline{2\,|\,17367}$$
$$43560\,|\,\overline{8683.5} = \text{Area in sq.ft.}$$
$$0.1993 = \text{Area in acres}$$

FIGURE 6.15. Computation of area.

Areas with Irregular Boundaries

To compute the area when one or more of the boundaries is an irregular line, a straight line is first run as close as possible to the irregular boundary and offsets are measured from this line. To simplify computations, the offsets should be taken at the same distance apart. Figure 6.16a shows a plot bounded by three straight sides and one irregular side. The area of the quadrangle *ABCD* may be computed by the method previously explained, and to this area is added the irregular area *EGKH* and the two triangles DEH and *GCK*. The area *EGKH* may be computed by adding together the series of trapezoids.

A more accurate method of finding the area of the irregular area of the plot is to apply *Simpson's one-third rule*, which is based on the assumption that the boundary curve is a series of parabolic curves. By this rule,

$$\text{area} = \frac{d}{3}\,(h_e + 2\Sigma h_{\text{odd}} + 4\Sigma h_{\text{even}} + h'_e)$$

in which d = distance between offsets. *This distance must be the same for all offsets.*

$$h_e = \text{length of the first offset (See Figure 6.16}b.)$$
$$2\Sigma h_{\text{odd}} = 2 \times \text{sum of the lengths of all the odd offsets}$$
$$4\Sigma h_{\text{even}} = 4 \times \text{sum of the lengths of all the even offsets}$$
$$h'_e = \text{length of the last offset (See Figure 6.16}b.)$$

This formula can only be used when there is an even number of strips. When there is an odd number of strips the area of the last strip is computed as a trapezoid, as illustrated in the following example.

Example. Compute the area of the irregular portion of the plot *DEFGC*, shown in Figure 6.16a, the dimensions shown being in feet.

Solution: Simpson's rule will be used. Note that an *even* number of strips will include only

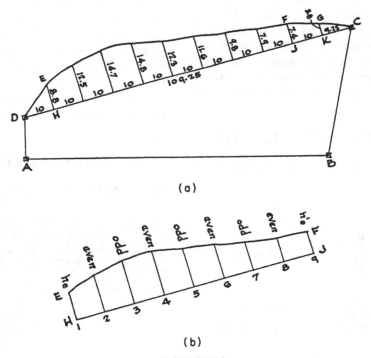

(a)

(b)

FIGURE 6.16.

the area *EFJH*. This area is again shown in Figure 6.16*b*, in which are shown the even- and odd-numbered offsets as well as the offsets designated as h_e and h'_e. Then

$$\text{area of } EFJH = \tfrac{10}{3}\,[8.8 + (2 \times 36.8)$$

$$+ (4 \times 46.8) + 7.4]$$

$$\text{area of } EFJH = \quad 923.3 \text{ ft}^2$$

area of trapezoid *FGKJ*

$$= \frac{7.4 + 3.8}{2} \times 10 = \quad 56.0$$

area of triangle *DEH*

$$= \frac{10 \times 8.8}{2} = \quad 44.0$$

area of triangle *GKC*

$$= \frac{9.25 \times 3.8}{2} = \quad 17.6$$

$$\text{area of } DEFGC = 1{,}040.9 \text{ ft}^2$$

Σh_{odd}	Σh_{even}
14.7	12.5
12.3	14.8
9.8	11.6
	7.9
36.8	46.8

When triangles which are not right triangles occur in the plot, their areas may be found by measuring the length of each of the three sides and using the formula:

$$\text{area} = \sqrt{s(s-a)(s-b)(s-c)}$$

in which *a*, *b*, and *c* are the lengths of the sides and $s = \tfrac{1}{2}(a + b + c)$. See Section 2.10.

To compute irregular areas, the traverse should be run as close to the irregular boundary line as possible in order to reduce the lengths of the offsets. When the boundary is a fairly smooth curve the intervals between offsets may be greater than when the curve is more irregular. Smaller intervals result in greater accuracy in the computed area.

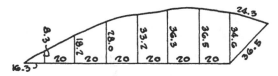

FIGURE 6.17.

PROBLEM 6.7.A.

Compute the area of land within the boundaries shown in Figure 6.17. Dimensions shown are in feet.

6.8 MISCELLANEOUS SURVEYING PROBLEMS

After the overall dimensions of the buildings to be placed on the plot have been determined, the architect is confronted with the problem of determining the distances of corners of the buildings, roadways, etc., from the boundary lines of the plot. If the intersecting boundary lines of the plot are not right angles, the best method of locating the corners of the building is by the use of coordinates. It is of great assistance to begin by drawing the plot accurately to scale. The survey plot received from the surveyor will show the lengths of the boundary lines and the various angles, but the coordinates of the corners are seldom shown. They should be computed for use in future computations. For the corners of the plot in our illustrative problem the coordinates were computed in Table 6.2 and are tabulated in Figure 6.15.

Finding Coordinates of the Corners of a Building

Figure 6.18 shows our property plotted with relation to the adjoining streets. The building we are to place on the plot is rectangular in plan, *FGHJ*, and has a length and width of 62 ft 5 in. and 28 ft $3\tfrac{1}{2}$ in., respectively. Zoning regulations require that the building must not be closer than 14 ft 0 in. to the street line *AB* or closer than 14 ft 0 in. to the party line *BC*,

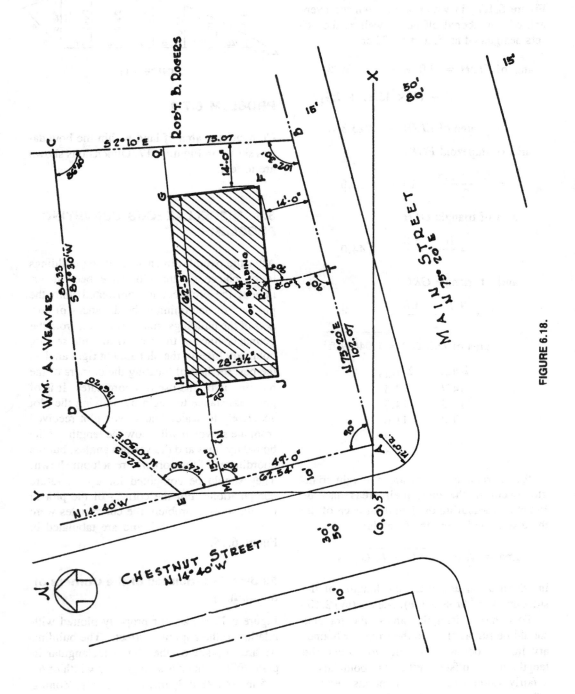

FIGURE 6.18.

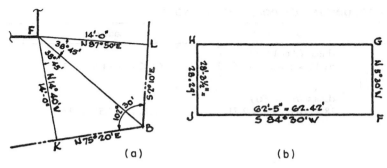

FIGURE 6.19.

The architect has decided to place the long axis of the building parallel to the north boundary line CD.

The corner of the building F in relation to the boundary lines AB and BC is shown in Figure 6.19a. Two equal right triangles, FLB and FKB, are formed. Since angles FBK and FBL each equal $\frac{1}{2} \times 102°\ 30'$, or $51°\ 15'$, angles BFK and BFL each equal $180° - (90 + 51°\ 15')$, or $38°\ 45'$. Then,

$$LB = KB = 14 \times \tan 38°\ 45' = 11.24'$$

The bearings of KB and BL are N $75°\ 20'$ E and S $2°\ 10'$ E, respectively; hence, since angles FKB and FLB are each $90°\ 0'$, the bearings of FK and FL are, respectively, N $14°\ 40'$ W and N $87°\ 50'$ E. We know the coordinates of point B to be $X_B = 114.63$ and $Y_B = 25.83$ (Figure 6.15); consequently we can now compute the coordinates of point F by the method explained in Section 6.4. These computations are shown in Table 6.4, and we find $X_F = 100.22$ and $Y_F = 36.53$.

Since GH is parallel to CD, (see Figure 6.18), the bearings of lines GH and $FJ = $ S $84°\ 30'$ W. Consequently, the building being rectangular, the bearings of JH and $FG = $ N $5°\ 30'$ W. See Figure 6.19b. The coordinates of point F having been determined, the coordinates of points G, H, and J can now be found, the computations being shown in Table 6.5. From these computations we find

$$X_F = 100.22 \quad \text{and} \quad Y_F = 36.53$$

$$X_G = 97.51 \quad \text{and} \quad Y_G = 64.69$$

$$X_H = 35.38 \quad \text{and} \quad Y_H = 58.71$$

$$X_J = 38.09 \quad \text{and} \quad Y_J = 30.55$$

Length and Bearing of a Missing Line

When the coordinates of the two ends of a line are known, the length and bearing of the line may be found by the methods explained in Section 6.4.

TABLE 6.4 COORDINATES OF POINT F

X Coordinates	Departures	Latitudes	Y Coordinates
$X_B = 114.63$ -10.87 $X_K = 103.76$	$11.24\ (\sin 75.33°) = 10.87$	$11.24\ (\cos 75.33°) = 2.845$	$Y_B = 25.83$ -2.84 $Y_K = 22.99$
	$14\ (\sin 14.67°) = 3.54$	$14\ (\cos 14.67°) = 13.54$	
-3.54 $X_F = 100.22$			$+13.54$ $Y_F = 36.53$

TABLE 6.5   COORDINATES OF POINTS G, H, AND J

X Coordinates	Departures	Latitudes	Y Coordinates
$X_F = 100.22$	(Lines *FJ* and *GH*)		$Y_F = 36.53$
−62.13	62.42 (sin 84.5°) = 62.13	62.42 (cos 84.5°) = 5.98	−5.98
$X_J = 38.09$			$Y_J = 30.55$
	(Lines *JH* and *FG*)		
−2.71	28.29 (sin 5.5°) = 2.71	28.29 (cos 5.5°) = 28.16	+28.16
$X_H = 35.38$			$Y_H = 58.71$
+62.13			+5.98
$X_G = 97.51$			$Y_G = 64.69$
+2.71			−28.16
$X_F = 100.22$			$Y_F = 36.53$

Example 1. It is required to build a fence between points *D* and *H*, Figure 6.18. What is the length of the line *DH* and what are the angles *EDH* and *HDC*?

Solution: The first step is to determine the coordinates of points *D* and *H*. This has already been done, and from Figure 6.15 and the preceding work we see that

$$X_D = 27.86 \quad \text{and} \quad Y_D = 92.75$$

$$X_H = 35.38 \quad \text{and} \quad Y_H = 58.71$$

Referring to the right triangle shown in Figure 6.20*a*, it is seen that

$$\text{distance } MH = X_H - X_D = 35.38$$

$$-27.86 = 7.52$$

$$\text{distance } DM = Y_D - Y_H = 92.75$$

$$-58.71 = 34.04$$

Then distance $DH = \sqrt{7.52^2 + 34.04^2}$

$$= \sqrt{56.55 + 1{,}158.72}$$

$$\text{distance } DH = \sqrt{1{,}215.27}$$

$$= 34.86 \text{ ft}$$

tan bearing of *DH*

$$(\text{Figure 6.20}a) = \frac{MH}{DM} = \frac{7.52}{34.04}$$

(see Section 6.4)

bearing of $DH = \text{S } 12° \, 25' \text{ E}$

Note that the bearing is taken to the nearest 05′, to be consistent with the accuracy of the survey.

Now that the bearings of the three intersecting lines *ED*, *HD*, and *CD* are known, the angles are readily established. Figure 6.20*b* shows the angles that lines *ED* and *HD* make with the north meridian. Thus, it is seen that the angle between *ED* and *HD* is 40° 50′ +

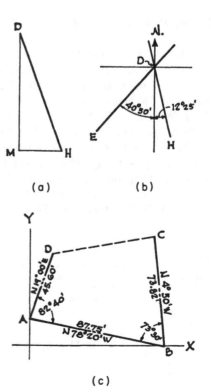

(a)

(b)

(c)

FIGURE 6.20.

12° 25', or angle $EDH = 53° 15'$. Since angle CDE is 136° 20' (see Figure 6.18), angle $CDH = 136° 20' - 53° 15'$, or 83° 0.5'.

Example 2. In surveying the plot $ABCD$ shown in Figure 6.20c, obstructions prevented measuring the interior angles at D and C and also the length of line DC. With the data shown in the figure, compute the length of DC and the angles ADC and DCB.

Solution:

STEP 1. First determine the coordinates of points D and C. Since the bearings and lengths of DA, AB, and BC have been established, the coordinates of points D and C may be found by the methods previously explained. The necessary computations are shown in Table 6.6. Note that the coordinates of D and C are

$$X_D = 14.85 \quad \text{and} \quad Y_D = 60.86$$

$$X_C = 79.72 \quad \text{and} \quad Y_C = 73.56$$

STEP 2. Compute the length of DC. Now that the coordinates of D and C have been found, the right triangle, shown in Figure 6.21a, shows that

departure $DC' = 79.72 - 14.85 = 64.87$

latitude $CC' = 73.56 - 60.86 = 12.70$

Hence, length of $DC = \sqrt{64.87^2 + 12.70^2} = \sqrt{4,369}$ and 66.10 ft

STEP 3. Compute the bearing of DC.

$$\tan \text{bearing of } DC = \frac{\text{departure}}{\text{latitude}} = \frac{64.87}{12.70}$$

(see Section 6.4)

bearing of $DC = $ N 78° 55' E

STEP 4. Compute angles ADC and DCB. We now know the bearings of lines AD, DC, and CB and, by the method explained in Section 6.2, we find angle $ADC = $ 120° 05' and angle $ADC = 120° 05'$ and angle $DCB = 83° 45'$. To check these angles we can use the rule given in Section 6.3. The sum of the interior angles $= 180(n - 2) = 180(4 - 2) = 360°$. Thus 73° 30' + 82° 40' + 120° 05' + 83° 45' = 360° 0'.

PROBLEM 6.8.A

In the surveys shown in Figures 6.21b and c, compute the magnitudes of the missing interior angles and the bearings and lengths of the missing sides.

Missing Line Problems

Many problems relating to a missing line may be solved by first finding the coordinates of the ends of the line.

TABLE 6.6 COORDINATES OF POINTS C AND D

X Coordinates	Departures	Latitudes	Y Coordinates
$X_A = 0$	(Line AB) 87.75 (sin 78.33°) = 85.94	87.75 (cos 78.33°) = 17.74	$Y_A = 17.74$
$\dfrac{+85.94}{X_B = 85.94}$			$\dfrac{-17.74}{Y_B = 0}$
	(Line BC) 73.82 (sin 4.83°) = 6.22	73.82 (cos 4.83°) = 73.56	
$\dfrac{-6.22}{X_C = 79.72}$			$\dfrac{+73.56}{Y_C = 73.56}$
	(Line DA) 45.60 (sin 19°) = 14.85	45.60 (cos 19°) = 43.12	
$\dfrac{+14.85}{X_D = 14.85}$			$\dfrac{+43.12}{Y_D = 60.86}$

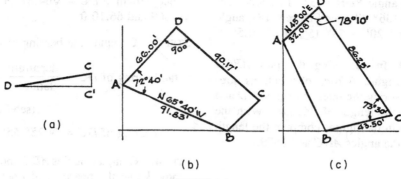

FIGURE 6.21.

Example. In Figure 6.18 line *MNP* represents the center line of a pathway. Point *M* is 49 ft 0 in. from point *A*, *MN* is 15 ft 0 in. in length. Determine the length of *NP* and the point at which the center of the pathway intersects the building, the length *HP*.

Solution:

STEP 1. Determine the coordinates of point *H*. These coordinates were established in Table 6.5. We found that

$$X_H = 35.38 \quad \text{and} \quad Y_H = 58.71$$

STEP 2. Determine the coordinates of point *N*. We know the bearing of line *AE* to be N 14° 40′ W, and, since *MN* makes an angle of 90° 0′ with *AE*, the bearing of *MN* = 90° 0′ − 14° 40′ = N 75° 20′ E. In Figure 6.15 we find that $X_A = 15.88$ and $Y_A = 0$. With these data we can now compute the coordinates of point *N*. This is shown in

Table 6.7, and we find that $X_N = 17.98$ and $Y_N = 51.20$.

STEP 3. Find the length and bearing of *NH*.

departure of $NH = 35.38 - 17.98$

$$= 17.40$$

latitude $NH = 58.71 - 51.20 = 7.51$

length of $NH = \sqrt{17.40^2 + 7.51^2}$

$$= \sqrt{359.2} = 18.95 \text{ ft}$$

$$\tan \text{bearing } NH = \frac{\text{departure}}{\text{latitude}} = \frac{17.40}{7.51}$$

(see Section 6.4)

bearing of $NH = $ N 66° 40′ E

STEP 4. Find the interior angle at *H* shown in the right triangle, Figure 6.22*a*. The lines *JH* and *FG* shown in Figure 6.19*b* are parallel; hence the bearing of *JH* and *PH* (see

TABLE 6.7 COORDINATES OF POINTS *M* AND *N*

X Coordinates	Departures	Latitudes	Y Coordinates
$X_A = 15.88$	(Line *AM*)		$Y_A = 0$
	49 (sin 14.67°) = 12.41	49 (cos 14.67°) = 47.40	
-12.41			$+47.40$
$X_M = 3.47$	(Line *MN*)		$Y_M = 47.40$
	15 (sin 75.33°) = 14.51	15 (cos 75.33°) = 3.80	
$+14.51$			$+3.80$
$X_N = 17.98$			$Y_N = 51.20$

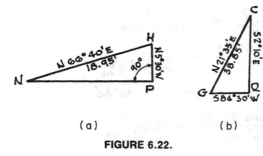

(a)

(b)

FIGURE 6.22.

Figure 6.22a) is N 5° 30' W. Therefore, angle $NHP = 66° 40' + 5° 30' = 72° 10'$.

STEP 5. Find HP and NP by solving the right triangle NHP, Figure 6.22a.

$$HP = 18.95 \times \cos 72° 10'$$

$$= 5.80 \text{ ft} = 5 \text{ ft } 9\tfrac{5}{8} \text{ in.}$$

$$NP = 18.95 \times \sin 72° 10'$$

$$= 18.04 \text{ ft} = 18 \text{ ft } 0\tfrac{1}{2} \text{ in.}$$

Example. It is desired to build a fence from G to Q shown on Figure 6.18. The fence is a continuation of the building line HG. Compute the length of the fence GQ and the distance CQ.

Solution:

STEP 1. Compute the length and bearing of the line GC.

$$X_C = 111.80 \text{ and} \qquad Y_C = 100.84$$

(from Figure 6.15)

$$X_G = 97.51 \text{ and} \qquad Y_G = 64.69$$

(from Figure 6.5)

departure   latitude
 of $GC = 14.29$ of $GC = 36.15$

$$\text{length of } GC = \sqrt{14.29^2 + 36.15^2}$$

$$= \sqrt{1,509} = 38.85 \text{ ft}$$

$$\tan \text{ bearing } GC = \frac{\text{departure}}{\text{latitude}} = \frac{14.29}{36.15}$$

(see Section 6.4)

$$\text{bearing } GC = \text{N } 21° 35' \text{ E}$$

STEP 2. Compute angles at G, C, and Q shown in Figure 6.22b. Since the bearings of the three sides of the triangle are known,

angle CGQ

$$= 84° 30' - 21° 35' \qquad = 62° 55'$$

angle QCG

$$= 21° 35' + 2° 10' \qquad = 23° 55'$$

angle GQC

$$= 180° - 2° 10' - 84° 30' = \underline{93° 20'}$$

$$180° 00'$$

STEP 3. Compute the lengths of lines CQ and GQ. These lengths are most readily found by the application of the sine law given in Section 2.9. This law states that in any triangle the sides are proportional to the sines of the opposite angles. The computations for these lengths are shown in Figure 6.23, and it is found that $CQ = 34.65$ ft and $GQ = 15.67$ ft.

PROBLEM 6.8.B

The line RST in Figure 6.18 represents the center line of an entrance pathway. If the distance RS is 8 ft 0 in., compute the lengths of ST and TB.

Areas of Plots

It is frequently necessary to compute the magnitudes of certain areas within the boundaries

$$\frac{38.85}{\sin 93°20'} = \frac{CQ}{\sin 62°55'} = \frac{CQ}{\sin 23°45'}$$

$$\text{Length } CQ = \frac{38.85 \times \sin 62°55'}{\sin 93°20'} = 34.65'$$
$$\text{or } 34'\text{-}7\tfrac{3}{4}''$$

Note: $\sin 93°20' = -\sin(180° - 93°20') = -\sin 86°40'$

$$\text{Length } GQ = \frac{38.85 \times \sin 23°45'}{\sin 93°20'} = 15.67'$$
$$\text{or } 15'\text{-}8''$$

FIGURE 6.23. Computation of line lengths.

TABLE 6.8 COORDINATES OF POINT Q

X Coordinates	Departures	Latitudes	Y Coordinates
$X_G = 97.51$	15.67 (sin 84.5°) = 15.60	15.67 (cos 84.5°) = 1.50	$Y_G = 64.69$
$+15.60$			$+1.50$
$X_Q = 113.11$			$Y_Q = 66.19$

COR.	COORDINATES		Y-DIFFERENCE		AREAS	
	X	Y	CORNER	VALUE	+	−
H	35.38	58.71	Q−D	−26.56		940
Q	113.11	66.19	C−H	+42.13	4765	
C	111.80	100.84	D−Q	+26.56	2969	
D	27.86	92.75	H−C	−42.13		1174

$$+7734 - 2114$$
$$-2114$$
$$\div 2 \overline{\,5620\,}$$
$$\div 9 \overline{\,2810\,} = \text{Area in sq. ft.}$$
$$312.2 = \text{Area in sq. yds.}$$

FIGURE 6.24. Computation of areas.

of a plot. These areas may often be computed by first dividing them into simple figures and then finding their sum. When, however, the coordinates of the various points of the survey have been previously found it may be convenient to use the system of computation given in Section 6.7.

Example. It is required to cover the area HQCD (see Figure 6.18) with bituminous paving. How many square yards of paving are required?

Solution: The coordinates of points C and D are found in Table 6.2, and the coordinates of point H are given in Table 6.5. It remains, therefore, to compute the coordinates of point Q. From Figure 6.23 it was found that GQ

has a length of 15.67 ft; its bearing is S 84° 30' W. The computations are shown in Table 6.8; $X_Q = 113.11$ and $Y_Q = 66.19$.

With these data we can use the formula given in Section 6.7. The simplest method of using the formula is to tabulate the terms; this is shown in Figure 6.24, and we find that area HQCD contains 312.2 square yards.

PROBLEM 6.8.C.

Compute the area of HQCD Figure 6.18, by considering it to be a trapezoid. Note that GQ was computed to be 15.67 ft. (Suggestion: Compute the perpendicular distance between the parallel lines DC and HQ.)

7

CIRCULAR HORIZONTAL CURVES

7.1 CIRCULAR CURVES

The curves commonly employed in building operations are *horizontal circular curves*, arcs of circles. Figure 7.1 shows two nonparallel lines, *A-P.C. and B-P.T.*, that intersect at point *V*, the *vertex*. A circular curve, whose radius is *R* and whose center is point *O*, connects the two straight lines. Point *P.C.*, the *point of curvature*, is the point at which the curve is tangent to the line *AV* and at which the circular curve begins. Point *P.T.*, the *point of tangency*, is the point at which the curve is tangent to the line *BV* and at which the straight line *P.T.-B* begins. The straight lines joining the curve are called *tangents*. Since the curve is tangent to lines *AV* and *BV* at points *P.C.* and *P.T.*, the angles *A-P.C.-O* and *B-P.T.-O* are right angles. The angle between the two radii at point *O* is angle *I*, the *included angle*. *I* is also always equal to the external angle at point *V*. The distances from *V* to *P.C.* and from *V* to *P.T.* are equal; this distance is *T*, the *tangent distance*.

If we know *R*, the radius of the arc, and also *I*, the included angle, the tangent distance *T* may be computed by the formula

$$T = R \tan \frac{I}{2}$$

The straight line connecting points *P.C.* and *P.T.* is *C*, the *chord*. Its length may be computed by the formula

$$C = 2R \sin \frac{I}{2}$$

7.2 LENGTH OF CURVES

To designate a specific curve, three elements should be given. They are

(a) the radius of the curve *R*;
(b) the included angle *I*;
(c) the length of the arc.

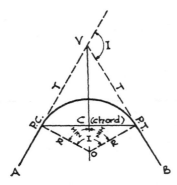

FIGURE 7.1.

FIGURE 7.2.

Whereas any two of these elements identify the curve, all three should be computed and shown on the drawings.

When the radius of the curve and the included angle are known, the length of the arc may be computed by two different methods. We know the *circumference* of a circle to be $2\pi R$ and that a circle contains 360°. Thus, to find the length of an arc, we simply consider it to be a certain part of a circumference.

Example 1: Consider the radius of the curve, shown in Figure 7.1 to be 40 ft 0 in. and that *I*, the included angle, is 85°20′. Compute the length of the curve from point *P.C.* to point *P.T.*

Solution: Since 85°20′ may be written 85.33°,

$$\frac{85.33}{360} \times 2\pi R = \frac{85.33}{360} \times 2 \times 3.1416 \times 40$$

$$= 59.57 \text{ ft the length of}$$

$$\text{the curve}$$

Now that the length of the curve has been found, all three elements of the curve are known and this information is shown on the drawing as indicated in Fig. 7.2. If the vertex and points of curvature or tangency are tied in by dimensions with other established points on the plot, no other information is needed for staking out the curve.

Example 2: Two intersecting lines and their bearings are shown in Figure 7.3*a*. A circular curve, whose radius is 50 ft 0 in., is to be constructed tangent to these lines. Compute the data required to dimension the curve completely.

Solution: The first step in the solution is to find *I*, the included angle. Since we know that *I* is also the magnitude of the external angle at the intersection of the two lines (see Figures 7.1 and 7.3*b*) and as the bearings of the lines are given, angle *I* can be computed by means of the principles explained in Section 6.2. Thus, by making a drawing showing the bearings of the lines, we see that 76°10′ − 35°40′ = 40°30′, the angle *I*. Now that *I* is known, we may find the length of the curve. Therefore,

$$\frac{40.5}{360} \times 2\pi R = \frac{40.5}{360} \times 2\pi \times 50$$

$$= 35.34 \text{ ft the length of}$$

$$\text{the curve}$$

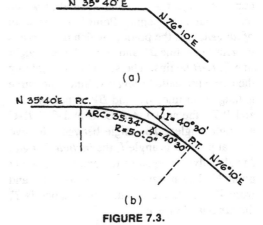

(a)

(b)

FIGURE 7.3.

The complete curve dimensions are shown in Figure 7.3*b*.

This example is the type of problem commonly encountered. A variant consists in having as data the length of the curve and the included angle, the radius being unknown.

Example: The included angle of a circular curve tangent to two straight lines is 156.59°, and the length of the curve is 62.28 ft. Compute the radius.

Solution:

$$\frac{156.59}{360} \times 2\pi R = 62.68 \text{ ft}$$

$$R = \frac{62.68}{2\pi} \times \frac{360}{156.59} = 22.93 \text{ ft}$$

PROBLEM 7.2.A. THROUGH D.

Each diagram in Figure 7.4 gives the bearings of two intersecting lines and the radius of the tangent circular curves. Compute the lengths of the arcs.

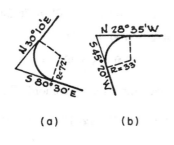

(a)   (b)

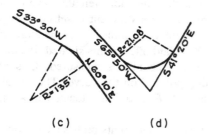

(c)   (d)

FIGURE 7.4.

PROBLEM 7.2.E THROUGH H.

Compute the radii for the given arc lengths and included angles.

	Length of Arc (ft)	Included Angle (degrees)
7.2.E	93.46	130.55
7.2.F	32.21	47.504
7.2.G	86.33	126.38
7.2.H.	22.64	33.27

7.3 LAYING OUT CIRCULAR CURVES

In laying out curves in the field, stakes are driven on the curve at a sufficient number of points to mark its exact location. Curves having comparatively small radii require a greater number of stakes than curves in which the radii are large. The method commonly employed in computing the location of the points on the curve is the *deflection angle method.*

7.4 DEFLECTION ANGLES

With respect to curves, a *deflection angle* is the angle that a chord makes with a tangent. Consider any point, such as *P*, on curve *AB* shown in Figure 7.5*a*. The deflection angle of point *P* is angle *VAP*. *The deflection angle is always one-half the angle subtended by the chord.* In Figure 7.5*a*, angle VAP = ½ angle POA.

To stake out a curve, the deflection angles and chord lengths are computed for a number of points on the curve. Then *P.C.*, the point of curvature, *P.T.*, the point of tangency and *V*, the vertex, are all located on the plot and marked with stakes. The transit is set up over the point of curvature (or point of tangency) and sighted on *V*, the vertex. The deflection angles are then turned off, and the chord distances are measured along these lines, thus determining the positions of the stakes on the curve. Generally a number of equal arcs are

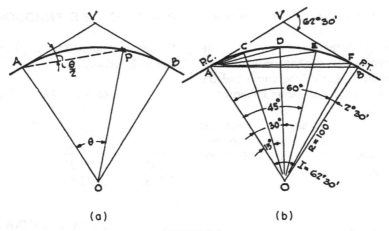

(a)   (b)

FIGURE 7.5.

laid off with a shorter arc remaining at the end of the curve, as shown in Figure 7.5b. The procedure of computing the deflection angles and chord lengths is illustrated in the following example.

Example: A curve having a radius of 100 ft 0 in. is tangent to two intersecting lines for which the external angle at their point of intersection, the included angle, is 62°30′, as shown in Figure 7.5b. Compute the location of four points on the curve.

Solution: Since the external angle at V is 62°30′, this is also I, the included angle at point O, the center of the circle of which AB is the arc. A line is now drawn from point O to point A (this point is also $P.C.$) and from this line we construct angles of 15°, 30°, 45°, and 60°, the lines of these angles intersecting the curve at ponts, C, D, E, and F, respectively.

In Section 7.1 we found that the length of a chord may be found by the formula

$$C \text{ (the chord)} = 2R \sin \frac{I}{2}$$

This formula, coupled with the fact that the deflection angle is one-half the angle subtended by the chord, enables us to compute the various chord lengths.

In Figure 7.5b chord AC subtends an angle of 15°0′ and therefore deflection angle VAC = one-half 15°0′, or 7°30′. Now, since

$$C = 2R \sin \frac{I}{2},$$

$$AC = 2 \times 100 \times \sin 7°30'$$

$$= 26.11 \text{ ft} \quad \text{the length of chord } AC$$

Similarly,

$$AD = 2 \times 100 \times \sin 15°0' = 51.76 \text{ ft}$$

$$AE = 2 \times 100 \times 22°30' = 76.54 \text{ ft}$$

$$AF = 2 \times 100 \times 30°0'$$

$$= 100.0 \text{ ft} \quad (\sin 30° = 0.5)$$

To check the remaining chord FB,

$$FB = 2 \times 100 \sin 1°15' = 4.36 \text{ ft}$$

Now that the chord lengths have been determined, the instrument is set up over point $P.C.$, sighted on point V, and the angle VAC, 7°30′, is laid off. On this sighted line the distance AC, 26.11′, is measured, thus establishing point C on the curve. The remaining points, D, E, and F are located in a similar manner.

In laying out points on the curve, obstructions may obscure the lines of sight. For example, in Figure 7.5*b* a tree or other obstacle might prevent points *E* and *F* from being visible from *P.C.* For such a condition points *C* and *D* could be located as previously described and the points *E* and *F* located from *P.T.*

It is sometimes convenient to divide the curve into a number of equal parts. This procedure is followed when the resulting deflection angles are not beyond the limitations of the surveying instrument.

PROBLEM 7.4.A.

For the curve shown in Figure 7.5*b*, four stakes are to be set so that arc *AB* is divided into five equal parts. Compute the various deflection angles and the chord distances.

7.5 LAYING OUT A GIVEN ARC DISTANCE

In order to locate full stations or other exact points on a curve, it is sometimes necessary to lay out a given arc distance from point *P.C.* Suppose, for example, the *arc* distance *P.C.-P*, in Figure 7.6, is given and we are required to compute the *chord* distance *P.C.-P*. To accomplish this, we first find the subtended angle of the arc *P.C.-P*; this is angle *POA*. We know that the deflection angle *VAP* is one-half angle *POA*, and this infor-

mation enables us to compute the chord distance *P.C.-P*.

Example: In the curve shown in Figure 7.6, the *arc* distance is 60 ft 0 in. from point *P.C.* to point *P* and the radius of the curve is 100 ft 0 in. Compute the deflection angle of point *P* and also the chord distance, *P.C.-P*.

Solution: The first step is to find the angle subtended by the arc of 60 ft 0 in.

$$\frac{\text{Angle}}{360} \times 2\pi R = 60$$

$$\text{Angle} = \frac{60 \times 360}{2\pi(100)} = 34.38°$$

Therefore, the deflection angle is one-half of this or 17.19°. The chord length is then determined as

$$\text{chord} = 2 \times R \times \sin\frac{I}{2}$$

$$= 2(100)\,(0.2955)$$

$$= 59.11 \text{ ft}$$

Now that both the deflection angle and chord distance have been computed, the instrument is set up over *P.C.* and sighted on *V*. The deflection angle *VAP* is turned off as close to 17.19° as the instrument will permit, and point *P* is staked out at 59.11 ft from point *P.C.*

PROBLEM 7.5.A.

On the curve shown in Figure 7.7*a*, it is desired to locate a point at an arc distance of 53.27 ft from point *P.C.* Compute the deflection angle and the chord distance.

PROBLEM 7.5.B.

On the curve shown in Figure 7.7*b*, it is desired to locate a point at an arc distance of 153

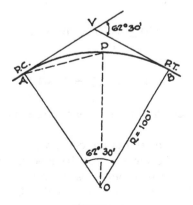

FIGURE 7.6.

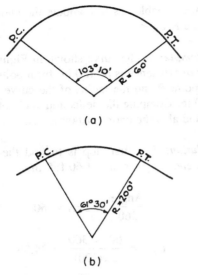

(a)

(b)

FIGURE 7.7.

ft from point *P.C.* Compute the deflection angle and the chord distance.

7.6 MIDDLE ORDINATES OF CIRCULAR CURVES

When two points on a circular curve have been located and the chord distance between them is known, other points between them on the curve are found without difficulty. In Figure 7.8, *A* and *B* are two points on a circular curve the radius of the which is *R*. The chord distance may be computed as explained in Section 7.1 or found by direct measurement in the field. Point *m* is located at the mid-point of chord *AB*, and line *mm'* is set off at right angles to *AB*; *mm'* is called the *middle ordinate* and its length is readily computed.

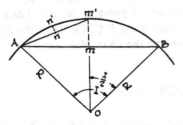

FIGURE 7.8.

In Section 7.1 we found the formula for computing the chord distance to be:

$$\text{chord distance} = 2 \times R \times \sin \frac{I}{2};$$

therefore,

$$\sin \frac{I}{2} = \frac{\frac{1}{2} \text{ chord distance}}{R} \qquad (1)$$

The length of the middle ordinate *mm'* may be found by the formula

$$mm' = R - R \cos \frac{I}{2} \qquad (2)$$

Example 1: In Figure 7.8 the radius of the circular curve is 120 ft 0 in. and the chord distance *AB* is 173.2 ft. Compute the length of *mm'*, the middle ordinate.

Solution: The first step is to find angle *I*/2. From Equation (1),

$$\sin \frac{I}{2} = \frac{86.6}{120}$$

therefore,

$$\frac{I}{2} = 46.19°$$

To find the length of *mm'*, we use Equation (2). Then

$$mm' = 120 - (120 \times \cos 46.19°)$$

and

$$mm' = 120 - 83.07 = 36.93 \text{ ft}$$

the length of the middle ordinate *mm'*

In order to lay out the curve with greater accuracy, the process may be repeated and successive middle ordinates, such as *nn'* in Figure 7.8, may be found. Usually, the subtended angle *I* will not be as large as that given in the previous example. *When the subtended angle I is less than 20°,* the following rule for finding the length of the middle ordinate gives lengths that are approximately correct and sufficiently accurate for most conditions:

$$\text{middle ordinate} = \frac{(\text{chord distance})^2}{8 \times R} \quad (3)$$

Example 2: In the circular curve shown in Figure 7.5*b*, compute the length of the middle ordinate of chord *AC* by use of equation (3).

Solution: From the computations shown in Section 7.4 the length of chord *AC* was found to be 26.11 ft; its subtended angle is 15° and the radius of the curve is 100 ft 0 in. From Equation (3),

$$\text{middle ordinate} = \frac{(26.11)^2}{8 \times 100}$$

$$= 0.852 \text{ ft the approximate length}$$

This figure was computed to the third decimal place in order to compare it with 0.856 ft, the exact length of the middle ordinate found by the use of equation (2).

When successive middle ordinates such as *nn'* in Figure 7.8 are sought, the following rule gives results that are approximately correct,

$$nn' = \frac{\text{middle ordinate distance } (mm')}{4} \quad (4)$$

Example 3: In the last example the length of the middle ordinate for chord *AC* was computed to be 0.85 ft. By the rule expressed by equation (4), compute the length of a sub-middle ordinate.

Solution:

$$nn' = \frac{0.85}{4} = 0.21'$$

The approximations that result from the use of Equations (3) and (4) are of convenience

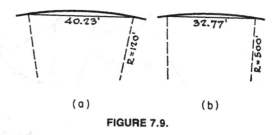

(a)   (b)

FIGURE 7.9.

to the draftsman when laying out curves of large radius.

PROBLEM 7.6.A. AND B.

For the chords of the circular curves shown in Figures 7.9*a* and *b*, compute the middle ordinates.

7.7 TANGENT CURVES

In addition to being tangent to straight lines, circular arcs are sometimes joined, the curves having a common tangent at the point at which they meet. Figure 7.10*a* shows the arcs of two circular curves both of which are tangent to the line V_1-V_2, the centers of the arcs being on the same side of the tangent line; the resulting curve is called a *compound curve.* Figure 7.10*b* shows the arcs of two circular curves having the common tangent V_1-V_2, the centers of the arcs being on opposite sides of the tangent; this type of curve is a *reverse curve.* In desiging roads and highways, reverse curves should be avoided whenever possible.

The method of establishing points on tangent curves is similar to the methods employed for simple curves. For conditions in which two curves meet at a common tangent, *a line joining the centers of the curves and the common point of tangency (line OO'P in Figure 7.10a and line OPO' in Figure 7.10b) is always a straight line and is always at right angles to the common tangent.*

In planning tangent curves the simplest procedure is to begin by drawing the tangent

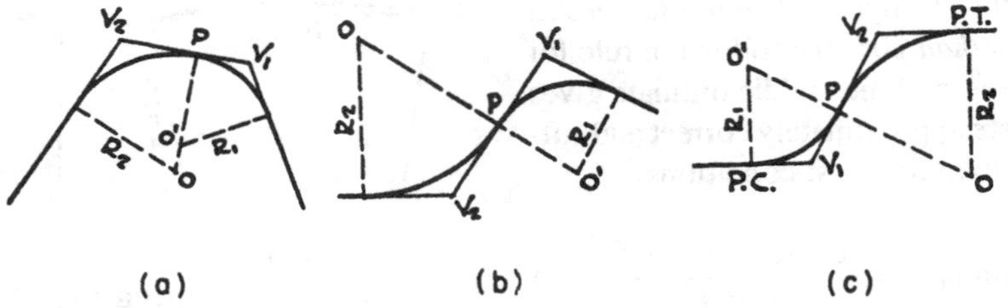

(a)   (b)   (c)

FIGURE 7.10.

line, V_1-V_2, computing its length, and selecting point P, the common point of tangency. After this, the tangent distances are made equal and the radii of the curves computed as previously described.

The method of connecting two straight lines with a reverse curve is shown in Figure 7.10c. A line V_1-V_2 intersecting the two straight lines is drawn and its length is computed; this line will be the tangent common to the two curves. A point, such as P, is se-

lected on line V_1-V_2, and through this point a line is drawn at right angles to V_1-V_2. The centers of the arcs of the curves will lie on this line. The tangent distances are now laid off on the two given lines; V_1-$P.C. = V_1$-P and V_2-$P.T. = V_2$-P. At points $P.C.$ and $P.T.$ lines are drawn at right angles to the two given lines. The intersection of these lines with the line through P perpendicular to V_1-V_2 determines points O' and O, the center of the arcs; thus the radii R_1 and R_2 are established.

8

LEVELLING

8.1 LEVELS

The *elevation* of a point is its vertical distance above or below a datum plane. *Levelling* is the process by means of which differences of elevation of two or more points are determined. The previous matter in this book has dealt with lines, angles, and areas assumed to be in the same plane. The surveyor, however, is also called upon to determine the rise and fall of the ground so that provision may be made for convenient access to buildings, grading, and proper drainage of the site.

The instruments used in levelling are the *level* and the *levelling rod*. Because of its accuracy in manufacture as well as the length (about 18 in.) of its telescope, the *engineers' level* affords the greatest accuracy in results. The engineers' transit and builders' level are also used for taking levels, but they do not afford the accuracy obtained by the engineers' level.

Levelling rods are about 7 ft in length and are extendable to twice their collapsed length.

The face of the rod is usually marked in feet and tenths of a foot, and the vernier on the movable target permits a reading of one hundredth of a foot. Some rods have graduations that give readings of one thousandth of a foot.

8.2 TAKING LEVELS

In order to determine the elevation of a point, it is necessary to begin with the elevation of some point that is either known or assumed.

Example: Figure 8.1 shows two points, *A* and *B*, the known elevation of point *A* being 202.58. Determine the elevation of point *B*.

Solution: STEP 1.   At any convenient point the instrument is set up and levelled. This point need not be on a line between points *A* and *B*.

STEP 2.   A reading is now taken on a rod held on point *A*. This reading is called a *back-sight* or *plus* (+) *sight*. The reading is 2.32,

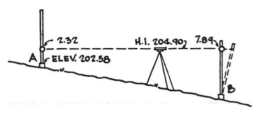

FIGURE 8.1.

hence the *height of the instrument* (H.I) is 202.58 + 2.32, or 204.90.

STEP 3.   Next, the instrument is revolved and is sighted on a rod held on point *B*; the reading is 7.89. This reading is the *foresight* or *minus* (−) *sight*. Since the height of the instrument is 204.90, 204.90 − 7.89 = 197.01, the elevation of point *B*.

The terms *backsight* and *foresight* are misleading since they do not indicate the direction in which the sights are taken. The backsight is the rod reading taken on a point whose elevation is known. The foresight is the rod reading on the point whose elevation is to be determined. The backsights are always plus, and the foresights are always minus.

In taking the levels of various points, the differences in level may be so great that the instrument must be moved from one position to another. When this is done, the instrument is sighted on a point whose elevation has been found previously. Such a point is called the *turning point*; before its elevation was found, its rod reading was a foresight. As an example, the elevation of point *B* (see Figure 8.1) has been found to be 197.01. For a new position of the instrument, point *B* is the turning point and becomes the backsight for determining the elevation of other points.

8.3   ACCURACY IN TAKING LEVELS

To determine elevations with accuracy, the instrument must be adjusted so that the bubble in the vial tube is in the exact center. In revolving the instrument the position of the bubble should be checked for each reading.

Regardless of the direction in which the instrument is sighted, the line of sight must be in a horizontal plane.

Another item of importance is to have the levelling rod held *in a vertical position when the reading is taken*. If the rod is not vertical, as indicated by the dotted line in Figure 8.1, a true reading cannot be taken. If the rod is not vertical the reading is always greater than the true reading.

8.4   DATUM AND BENCH MARK

The theoretical level plane to which elevations are referred is called *datum;* it is usually mean sea level. A *bench mark* is a permanent point whose elevation, above or below datum, is known. The U.S. Coast and Geodetic Survey and local authorities have established the elevations of numerous bench marks throughout the country, and from these points other bench marks may be established. The location and elevation of these official bench marks may be obtained from the engineering departments of the civil authorities. When showing elevations on a drawing, a note should be given to indicate the datum that is used: U.S. Geodetic or local. When no authoritative bench mark is available in the vicinity of the proposed work, any convenient permanent point may be taken as a datum and assigned an elevation, usually 100. This point is the height on which all other elevations shown on the drawing are based. This procedure, however, should be followed only when an established bench mark is not available.

During the construction of a building it is generally desirable to establish a temporary bench mark on the site. This bench mark is used to determine the various elevations shown on the drawing. In locating such a bench mark, care should be exercised to see that it is in a protected position and that it is not likely to be disturbed during construction operations. Points on nearby buildings, curbstones, or other permanent objects are preferable to wooden stakes used as bench marks.

8.5 ERRORS DUE TO CURVATURE AND REFRACTION

An error in levels is always present when the distance between the points is great. This error results from the curvature of the earth's surface and the refraction or bending of the line of sight as it passes through the atmosphere. In sights of ordinary length, the combined error is about 0.001 ft in a distance of 200 ft, so small that it is neglected in computations. In taking levels, *the error may be eliminated entirely by placing the instrument at equal distances from the backsight and foresight.* This is called *balancing the sights;* the equal distances may be approximated by the eye.

8.6 CORRECTING ERRORS IN LEVELS

In taking levels we begin with a bench mark of known or assumed elevation, preferably an *official bench mark.* Curb elevations, as shown on city or highway plans, should be used with caution since they may have settled after their installation. In order to check the accuracy of work, closed circuits returning to the original bench mark should be made whenever possible. In road surveys, sewer lines, etc., a closed circuit is impracticable. In such cases the levels should be tied in with other bench marks along the way, using them as turning points.

Figure 8.2 shows the various positions of the instrument obtaining the elevations for the corners of the plot shown in Figure 6.18. By consulting maps in the township engineer's office, the position of a bench mark was found to be on the south side of Main Street, as indicated in Figure 8.2. It was described as a stone marker, and the elevation given was 123.37. This bench mark was located at the site and the instrument set up at point 1, a point so located that the backsight on *BM* and the foresight on point *A* were approximately equal distances. Note that the instrument need

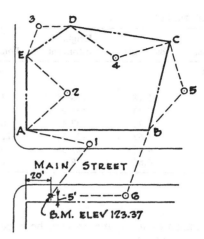

FIGURE 8.2.

not be set on a line between the backsight and foresight. Readings on the two points were then taken, and the instrument was moved to point 2. From this position readings were taken on point *A* (the turning point) as a backsight and point *E*, the foresight. The procedure was continued around the plot, using the corners of the plot as turning points, and returning to the original bench mark. The result of the readings were recorded as shown in Figure 8.3; the computed elevations are given in the fifth column.

Note that in returning to the bench mark, the computed elevation was found to be in error by 0.02 ft, (123.39 − 123.37). This error should be distributed among all the points in proportion to the distance travelled around the circuit. However, if this were done, the elevations would be given in thousandths of a foot. Such elevations would imply greater accuracy than could be obtained by the instrument. In this problem it is sufficiently accurate to drop the elevation of point *E* 0.01 ft and to drop point *C* an additional 0.01 ft, thus balancing the circuit. The corrected elevations are shown in the sixth column of the tabulation in Figure 8.3.

When levelling is done with an engineer's level the error in feet should not exceed $0.05 \sqrt{M}$, in which M is the length of the circuit in miles. When a less accurate instrument is used, such as an engineers' transit or build-

FIELD NOTES

Station	Backsight	Height of Instrument	Foresight	Elevation	Corrected Elevation	Remarks
B.M.	—	—	—	123.37	123.37	Stone marker in s. side of Main St.
A	5.18	128.55	3.98	124.57	124.57	
E	7.35	131.92	5.23	126.69	126.68	
D	10.78	137.47	6.71	130.76	130.75	
C	3.62	134.38	11.60	122.78	122.76	
B	4.80	127.58	6.32	121.26	121.24	
B.M.	6.85	128.11	4.72	123.39	123.37	

FIGURE 8.3.

ers' transit or level, the error to be expected will probably be two or three times this amount.

When the sight distances are less than 200 ft, the errors resulting from curvature and refraction are small and a saving of time may be effected by taking fewer turning points and taking several elevations from a single location of the instrument.

For example, suppose that in the preceding illustration we had decided to set up the instrument at point 2 (see Figure 8.2), from this point sight points A, E, and D, and then, using point D as a turning point, set up the instrument at point 4 and sight points C, B, and the bench mark. The field notes and computations would have resulted in the tabulation shown in Figure 8.4. In this instance the error of vertical closure is 0.03. To correct this discrepancy the height of the instrument is reduced 0.01 at point 2, 0.02 at point 4, and the

bench mark is reduced 0.03 to check with the original bench mark of 123.37.

8.7 INVERSE LEVELLING

When it is necessary to determine the elevation of the underside of a bridge, a ceiling, or similar object, the following method may be used.

Suppose that we are given the elevation of point A, in Figure 8.5, as 24.78 and are required to find the elevation of point B. The instrument is set up, and a reading is taken on the rod held on point A. This reading, 4.93, is a backsight or plus sight. The rod is now held on point B in an inverted position; that is, the zero mark on the rod is at the top. The reading is taken and is found to be 6.51. From the diagram it is seen that the elevation of point B is 24.78 + 4.93 + 6.51, or 36.22.

FIELD NOTES

Station	Backsight	Height of Instrument	Foresight	Elevation	Corrected Elevation
B.M.	—	—	—	123.37	123.37
2	8.30	131.67	—	—	131.66
A	—	—	7.10	124.57	124.56
E	—	—	4.98	126.69	126.68
D (t.p.)	—	—	0.91	130.76	130.75
4	3.62	134.38	—	—	134.36
C	—	—	11.60	122.78	122.76
B	—	—	13.02	121.36	121.34
B.M.	—	—	10.98	123.40	123.37

FIGURE 8.4.

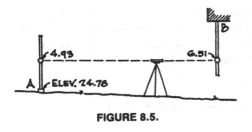

FIGURE 8.5.

8.8 PROFILES

Surveys of roads, railways, sewer lines, and other layouts of a linear character are often plotted in profile. A *profile* is the vertical projection of the line of intersection of a vertical plane with the surface of the ground. Profiles are usually taken at the center lines of roads and highways.

In plotting profiles it is customary to use an exaggerated vertical scale to show more clearly the changes in elevations. The horizontal scale is generally five or ten times the vertical scale. For example, if the horizontal scale is 1 in. = 40 ft 0 in., the vertical scale might be $\frac{1}{8}$ in. or even $\frac{1}{4}$ in. = 1 ft 0 in. Specifically prepared profile paper with ruled lines is of great convenience in plotting profiles. An example of a roadway profile is shown in Fig. 10.6.

Figure 8.6*a* shows the field notes, back-

FIELD NOTES

Station	Backsight	Height of Instrument	Foresight	Elevation
(a) Problem 8.8.A				
B.M.	3.42			197.78
A	4.35		2.89	
B	5.29		3.71	
C	1.71		5.47	
D	2.64		4.21	
B.M.			1.13	
(b) Problem 8.8.B				
B.M.	9.95			326.98
1				
A			3.26	
B (t.p.)	4.37		5.32	
2				
C			6.72	
D			4.58	
E (t.p.)	3.42		1.17	
3				
F			2.31	
B.M.			11.25	
(c) Problem 8.8.C				
BM_1	2.21			78.03
A	5.73		3.24	
B	3.86		8.43	
C	1.73		5.21	
BM_2	8.26		3.61	71.07
D	4.92		6.98	
E	7.08		1.56	
BM_3	6.50		9.15	73.64
F	2.61		7.76	
G	5.56		4.36	
H	1.20		3.87	
BM_4			5.01	68.51

FIGURE 8.6.

sights, foresights, and bench mark for a four-sided plot. The corners of the plot, A, B, C, and D, are used as turning points. See Problem 8.8.A.

The field notes for a six-sided plot are shown in Figure 8.6b. Points A, B, C, D, E, and F are the corners of the plot. Points 1, 2, and 3 are positions of the instrument. Points B and E are used as turning points. See Problem 8.8.B.

Figure 8.6c shows the field notes for the linear survey of a sewer line. Points A, B, C, D, E, F, G, and H are points along the route whose elevations are to be determined. These points, as well as the bench marks, BM_1, BM_2, and BM_3, are used as turning points. See Problem 8.8.C.

PROBLEM 8.8.A. Through C.

Figures 8.6A, B, C show field notes for certain problems in levelling. For each set of field notes the data balance out; there is no error of closure. Compute the various instrument heights and the site elevations of each point.

9

CONTOURS

A map is a two-dimensional element and various "tricks" must be used to represent the true three-dimensional form of the ground surface. The principal means for achieving this with line drawing is through the use of contours. This chapter explains the nature and use of contours and how they can be produced from survey data.

9.1 CONTOUR LINES

The usual map shows only two dimensions, length, and breadth. Various devices are employed to indicate the third dimension or relative differences in elevation, but the most practical method is the use of *contours*. Often, the differences in elevation of a site may be more thoroughly understood by inspecting a contour map than by inspecting the site itself.

A contour is a line drawn on a map or plan that connects all points having the same height above some reference plane. The reference plane is the *datum plane*, and on many maps

it is mean sea level. The vertical distance above the datum plane is the elevation. A contour may be visualized as the intersection (shown in plan) of a level plane, such as the surface of a body of water, with the undulating surface of the ground. Shore lines of bodies of still water illustrate contour lines; a rise or fall of water creates other contour lines. This is illustrated in Figure 9.1a, which shows a small island in both plan and elevation. As the water rises new shore lines are formed and these lines, in plan, constitute contours.

9.2 CONTOUR INTERVALS

A *contour interval* is the *vertical* distance between contours. A smaller contour interval will result in a greater number of contours on a map. The selection of the contour interval depends upon several factors: the purpose for which the map is to be used, the scale of the drawing, the roughness of the terrain, and the cost of obtaining the data required to plot the

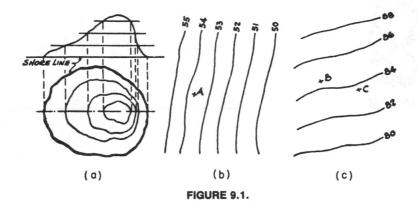

(a)   (b)   (c)

FIGURE 9.1.

contours. On small-scale maps, contour intervals of 50 ft and 100 ft are often used. For site plans, however, where more detailed information is demanded, intervals of 5 ft, 2 ft, and 1 ft are commonly employed. For building sites 1 ft intervals are recommended.

When the contour interval has been decided upon, the same interval should be maintained for the entire drawing. More than one contour interval on a drawing frequently leads to errors in interpretation. When a certain detail requires more information than is afforded by the contours shown, *intermediate contours* are sometimes drawn between the regular contours; they should be drawn with a light dotted or dashed line and should extend only as far as the detail requires.

9.3   SIGNIFICANCE AND USE OF CONTOURS

Map contours can be read for various purposes. The general ground surface form can be interpreted, but specific data can also be determined when needed.

Determining Elevations from Contours

On an accurately drawn contour plan, the elevation of any point may be determined by interpolation. The contour interval for the contours shown in Figure 9.1b is 1 ft, the

contours 50, 51, 52, 53, 54, and 55 being shown. Suppose that we are asked to determine the elevation of the point A. This point lies about 0.7 the distance from contour 53 to contour 54. Since each contour indicates a *vertical* distance of 1 ft, point A has an elevation of 53.7.

Figure 9.1c shows contours drawn with 2 ft intervals, as indicated by their numbers. To find the elevation of point B we first note that it is located 0.3 the distance from contour 84 to contour 86. As the contour interval is 2 ft, the elevation of point B is 84 + (0.3 × 2), or 84.6. Point C lies at 0.8 the distance from contour 82 to contour 84. The contour interval being 2 ft, the elevation of point C is 82 + (0.8 × 2), or 83.6.

General Significance of Contours

The contours of a map reveal definite characteristics of the terrain. A knowledge of these characteristics and their significance is essential in their interpretation.

1. Closely spaced contours at the higher elevations with greater spacings at lower levels indicate a concave slope. When the spacing is greater at the top of a slope and closer together at the bottom, the slope of the ground is convex. These conditions are indicated in Figures 9.2a and b, respectively.

2. Evenly spaced contours indicate a uniform slope. On a plane surface the contours are straight, evenly spaced, and parallel.

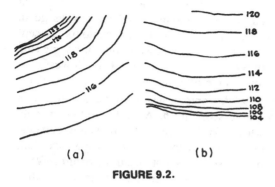

FIGURE 9.2.

3. Every contour is a continuous line which *closes upon itself* somewhere on the earth's surface though not necessarily within the limits of a drawing. A contour cannot stop within the confines of a drawing; it must be a closed curve or, if it enters at the border of a drawing, it must leave the sheet at some other point on the border. Figure 9.3a shows a plan of a pathway and steps with contour line 50 apparently stopping at the cheek wall of the steps. Figure 9.3b, a perspective, shows that this contour line actually follows around the cheeks and a riser of the steps.

4. A closed contour surrounded by other contours indicates either a summit or a depression, the distinction being indicated by the contour numbering. Since no numbering is shown on contours A and B, (see Figure 9.4a), the contours might indicate either a summit or a depression at the top of a slope.

If contours A and B were numbered 136 and 137, respectively, we would know that a summit was indicated. The numbering of the contours in Figure 9.4b indicates a depression at the bottom of a slope. To aid further in the identification of a depression, short lines drawn at right angles to a contour are sometimes used. These indications, called *hachures*, are shown at contour 133 in Figure 9.4b. The highest or lowest elevation is shown by a *spot elevation*, such as elevation 132.8 in Figure 9.4b.

5. Except for one condition, contour lines never cross each other, for such a condition would indicate a point having two different elevations. The exception is a vertical or overhanging cliff, as shown in both plan and profile in Figure 9.5a. For such a condition one contour must cross another at two points. Figure 9.5a shows the method of plotting a profile, the profile being taken on line A-A. Contours may appear to coincide at vertical excavations or at buildings. The perspective drawings, (see Figure 9.5b and c) show that the contours actually run around the face of the excavation or the face of a building.

6. Contour lines are perpendicular to lines of the steepest slope. Figure 9.5d shows two contour lines, 41 and 42. Since there is the same difference in elevation between point A and any point on contour 41, the steepest slope is found on the shortest line between the two contours. This line is a line at right angles

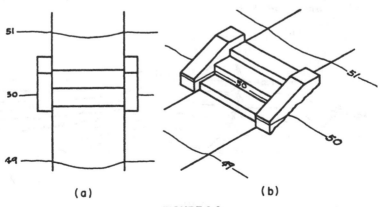

(a)   (b)

FIGURE 9.3.

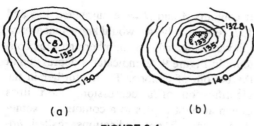

(a)

(b)

FIGURE 9.4.

to the contours. For the same reason, when contours cross ridge or valley lines they are perpendicular to them.

7. A stream and adjacent contour lines are shown in Figure 9.5e. When a contour line crosses a river or a stream, the contour first goes upstream, crosses the stream at right angles (the thread of the stream is the line of greatest slope), and then follows downstream.

8. The highest contours along ridges and the lowest contours in valleys always go in pairs. Figure 9.5f shows the contours adjacent to a stream. If the lowest contour on one side of the stream is 121, then A, the lowest contour on the opposite side, must also be 121.

9. A contour never splits, as shown in Figure 9.6a and b. A split could occur only when a knife-edged ridge or valley coincide exactly with a contour line. Such a condition, of course, does not occur in nature. When sharp ridges or depressions do occur they would probably be represented as indicated in Figures 9.6c and d.

9.4 PLOTTING CONTOURS

The best method of plotting contour lines for relatively small areas, such as building sites,

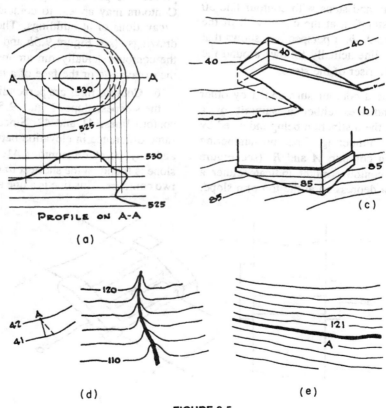

PROFILE ON A-A

(a)

(b)

(c)

(d)

(e)

FIGURE 9.5.

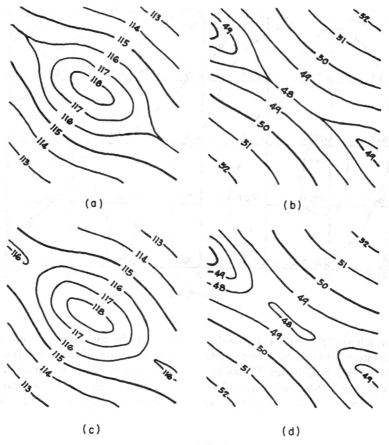

(a)

(b)

(c)

(d)

FIGURE 9.6.

is known as the cross-section or grid method. By the use of the transit and tape the plot of ground is first divided into a series of squares, called a grid; Figure 9.7 shows such a grid. For the purpose of identifying specific points on the plot, the horizontal grid lines are lettered A, B, C, D, and E, as shown. The vertical lines are numbered 1 to 7, inclusive. As an example, with this system of notation the center of this particular plot is identified as C-4. The corners of the grid squares are marked with temporary stakes; the elevations of the ground at these points are taken and are so marked on the plan, as shown on Figure 9.7.

The unevenness of the ground and the purpose for which the contour map is to be used determine the size of the grid squares; they vary from 10 ft to 100 ft.

Having laid out the grid and marked the elevations at the corners of the squares, our next task is to draw the contours. For the plot shown in Figure 9.7 we will use 1 ft intervals. We note that the highest and lowest contour lines will be 80 and 75, respectively. The only contour line that will cross the corner of a grid square is 77; it intersects point A-4. Hence, we must now find the remaining points on the grid lines at which the contour lines will cross.

As an example, the elevations of points E-5 and E-6 are 75.5 and 74.5, respectively. Obviously, contour line 75 will intersect grid line E at the point halfway between grid line 5 and 6. Next consider points D-5 and D-6, the el-

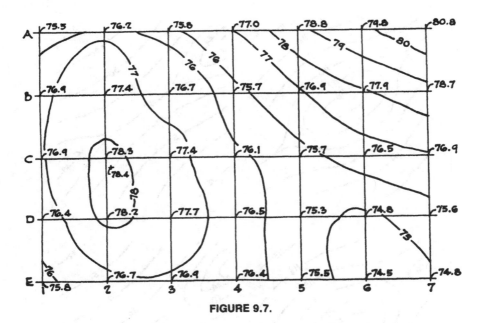

FIGURE 9.7.

evations of which are 75.3 and 74.8. The difference in elevation between these two points is 75.3 − 74.8, or 0.5. Contour 75 lies between points D-5 and D-6, and, since 75 − 74.8 = 0.2, contour 75 will cross grid line D

at two-fifths the distance from D-6 to D-5. This, of course, is interpolation. It may be done graphically, the underlying principle being the division of a line into any number of equal parts by the use of a scale. Figure

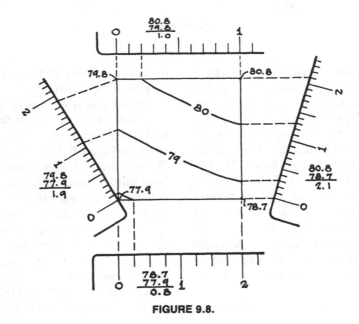

FIGURE 9.8.

9.8 is an enlarged drawing at the grid square shown in the upper right-hand corner of Figure 9.7. Any convenient scales are used, their purpose being to find the points of intersection of contour lines 78, 79, and 80 on the grid lines. Scales are laid adjacent to the grid square, and the figure indicates the procedure of finding the contour points. Minute accuracy, however is unnecessary, and an experienced topographer performs the interpolations mentally. When the contour points on all the grid lines have been determined the contour lines are drawn in *smooth* curves, as shown in Figure 9.7.

PROBLEM 9.4.A. AND B.

The tabulations that follow give the elevations of the grid intersections for two separate plot plans in a manner similar to that shown in Figure 9.7. Lay out grids using squares 1 in. in a side and draw the contour lines for one ft intervals.

9.5 GRADES

After the contour map of the site has been drawn, it can be used advantageously in determining the most desirable position for the buildings, roadways, paths, and required grading. Invariably, grading must be performed, and this will result in the revision of certain contour lines. A study of the natural contours and the contours showing new grades enables us to compute the volumes of cut and fill that are required. See Chapter 11.

With respect to site engineering, the term *grade* has two different meanings. Very often the word *grade* is used synonymously with the word *elevation*, the height of the earth's surface at some specific point.

Another meaning of the word grade is *gradient* or *slope*. If we are given two points on a plan, one higher than the other, the grade is found by dividing the difference in their ele-

Data for Problem 9.4.A
(Elevations in ft)

	1	2	3	4	5	6
A	35.3	36.4	37.0	38.1	38.9	39.6
B	36.9	37.9	38.8	39.9	40.4	39.7
C	38.0	39.0	40.4	41.5	40.4	39.0
D	38.8	40.3	41.7	40.4	39.2	38.3
E	39.3	40.5	39.9	39.2	38.3	37.3
F	39.5	39.4	38.9	38.0	37.1	36.3

Data for Problem 9.4.B
(Elevations in ft)

	1	2	3	4	5	6	7
A	95.0	94.8	95.7	95.6	95.5	94.8	94.1
B	95.7	94.7	96.4	97.2	97.3	96.3	95.0
C	96.7	95.2	95.5	97.4	98.6	97.4	95.7
D	97.5	96.4	94.8	95.8	96.8	96.7	95.8
E	97.1	98.1	96.4	94.9	95.3	95.4	94.6
F	96.4	98.0	97.8	96.6	94.9	93.0	93.2
G	95.4	96.6	97.4	96.8	95.6	94.4	92.6

vations by the horizontal difference (length) between them; that is,

$$G = D \div L \text{ or } D = G \times L \text{ or } L = D \div G$$

in which

G = the grade (gradient)

D = the difference in elevation between

the two points

L = the horizontal length between the

two points

The grade is expressed as a decimal or as a percentage. As an example, Figure 9.9 represents the plan of a roadway of uniform slope, the dot-and-dash line being its center line. Points A and B have elevations of 158.62 and 155.72, respectively, and the horizontal distance between them is 116 ft. The difference in elevation is 2.90 ft. Then $G = D/L$, or $G = 2.90/116 = 0.025$, or 2.5% of the grade (slope) of the roadway. A common indication of a grade is shown in Figure 9.9, with an arrow pointing down the slope and the grade expressed as a decimal.

When the grade of a line is known, the elevation of any point on the line may be determined by use of the formula $G = D \div L$. Suppose, for example, we wish to know the elevation of a point on line AB (see Figure 9.9) that is 22 ft to the right of point A. Then $D = G \times L$, or $D = 0.025 \times 22$, and $D = 0.55$, the difference in elevation. Hence, $158.62 - 0.55 = 158.07$ ft, the elevation of the point 22 ft to the right of point A. Similarly, we can find the elevation of a point 46 ft to the left of point B. $D = 0.025 \times 46 =$

1.15, and $155.72 + 1.15 = 156.87$, the elevation of the point.

A problem that frequently arises is to determine the point on a line that is intersected by a certain contour line. For example, find the point on line AB (see Figure 9.9) at which contour line 158 would cross. The difference in elevation between this point and point A is $158.62 - 158.00$, or 0.62 ft. Then $L = D \div G$, or $L = 0.62 \div 0.025 = 24.8$ ft, the distance to the right from point A to contour line 158. To find the distance to the right from contour 158 to contour 157 on line AB, $D = 1$ ft 0 in. and $L = 1.0 \div 0.025 = 40$ ft 0 in. Similarly, contour 156 will be $0.28 \div 0.025$, or 11.2 ft to the left of point B.

There are three terms in the basic formula $G = D \div L$. When any two are known, the third term may be computed. The method of computing grades and locating points that have been explained in this article must be thoroughly understood, for similar problems occur many times in the succeeding examples relating to contours.

Laying Out a Roadway of a Given Grade

Figure 9.10 shows the contour lines of a given site; the contour interval is 5 ft. Suppose we are required to determine the center line of a roadway connecting points A and B, whose maximum grade will not exceed 4%. Since the contour interval is 5 ft, the difference in elevation on the contour lines is 5 ft. Then $L = D \div G$, or $L = 5.00 \div 0.04$, or 125 ft, the minimum length of roadway between any two contour lines. Therefore, with point A as a center, lay off an arc having a radius of 125 ft to intersect contour 145; the intersection is point C. In the same manner, with point C as the center, find point D on contour 150, and with D as a center locate point E on contour 155. Note that if a straight line 125 ft in length is drawn between points E and G (on contour 160), a part of the line would be almost level and the remaining portion would have a grade

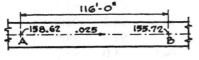

FIGURE 9.9.

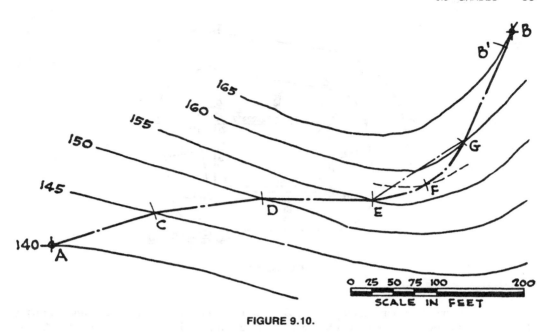

FIGURE 9.10.

exceeding the 4% maximum. For such a condition, a half-contour (interval 2.5 ft) is drawn between contours 155 and 160 and arcs *EF* and *FG* are laid off with radii of 125 ÷ 2, or 62.5 ft. A curve is then constructed connecting points *E*, *F*, and *G*. Since the arcs are longer than the straight lines between points *E*, *F*, and *G*, the grades will be slightly less than the maximum 4%. An arc with a 125 ft radius intersects contour 165 at point *B'*. As line *GB* is slightly longer than *GB'*, the grade of *GB* is somewhat less than the 4% maximum and thus the line indicating the center line of the roadway fulfills the requirements of the problem.

Various considerations may effect the final location of a roadway; the ideal solution is the roadway having the shortest length and requiring the least cut and fill. The roadway established in our solution started from point *A*. If we had selected point *B* as a starting point a slightly different route would have been obtained. For this type of problem widely divergent routes may fulfill all the requirements, but usually one solution is found to be more economical than the others.

Establishing Grades for Site Drainage

No natural site is exactly level, and it is not desirable that any portion of a finished (man-made) site be so. So-called level spaces, such as athletic fields, lawns, or play areas, must have a slight slope to permit rain water to run off. The earth surfaces adjacent to buildings should always be sloped so that surface water will drain *away from the building*.

Figure 9.11 shows a rectangular building around which the soil is to be graded to provide a minimum slope for the drainage of surface water. The building shown is 150 ft × 85 ft, and spot elevations of finished surface elevations are given at the corners of the building. These locations, indicated by crossed lines, are at the corners but are shown a slight distance from the building for the sake of legibility. The minimum grade (slope) of the earth surface will be assumed to be 2%. Our problem is to determine the contours of the graded soil that will provide this slope *at right angles to the faces of the building*.

We will begin at corner *A*, at which the finished elevation is to be 85.92, by drawing

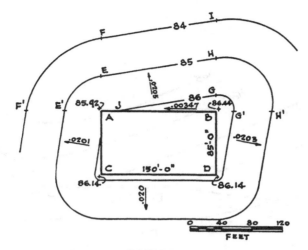

FIGURE 9.11.

a line from point A perpendicular to the AB face of the building; contour 85 will be somewhere on this line. The difference in elevation between the elevation at point A and contour 85 is $85.92 - 85$, or 0.92 ft. Then $L = D \div G$ (see Section 9.5), and $L = 0.92 \div 0.02$, or 46 ft. Hence, a point on the perpendicular line at 46 ft from point A is a point on contour 85; it is point E. In the same manner point E' is located 46 ft from the AC face of the building.

Next, let us find the point at which the line perpendicular to the AB face of the building intersects contour 84. The difference in elevation between contours 85 and 84 is 1 ft. Then $L = D \div G$, and $L = 1 \div 0.02$, or 50 ft, the distance from point E to contour line 84; this is point F. Similarly, 50 ft from E' locates point F'.

To find points G and G', $86.44 - 86 = 0.44$ ft. Then $L = D \div G$, or $L = 0.44 \div 0.02 = 22$ ft, the perpendicular distances from point B. Like the distance EF, distances, GH, $G'H'$, and HI are all 50 ft.

The same procedure is followed at corners C and D, and other points on the 86, 85, and 84 contours are thus found. Points having the same elevation are now joined to produce the contour line. For example, points E and H are connected with a straight line because the

earth surface between these points forms a level straight line. Contour line 86, GJ, is drawn parallel to EH through point G; it intersects the AB face of the building at point J. Distance AJ may be found as follows: The difference in elevation between points A and B is $86.44 - 85.92$, or 0.52 ft, and the horizontal distance between them is 150 ft. Then $G = D \div L$, or $G = 0.52 \div 150 = 0.00347$, the slope of the earth along line AB, as shown in Figure 9.11. The difference in elevation between points A and J is $86 - 85.92$, or 0.08 ft. Since $L = D \div G$, $L = 0.08 \div 0.00347$, or 23 ft, the length of AJ.

At the corners of the building, circular arcs, $F'F$ for example, are drawn to complete the contours. Although the layout work is done mechanically, the contours of ground areas on final drawings are drawn free hand. When contour lines cross paths or paved roadways, they are drawn mechanically.

Note that the earth surface in contact with the CD face of the building is level, points C and D each being 86.14. For this reason the contours are parallel to the face of the building and are 50 ft apart; this results in the minimum required slope, 0.02. On the other three faces of the building, however, the contours are not parallel to the face of the building and the slopes of the ground, *taken at right angles to the contours*, will be slightly in excess of the 2% minimum and will thus fulfill the requirements of the problem.

PROBLEM 9.5.A.

A rectangular building, 150 ft × 83 ft in plan, is indicated in Figure 9.12a; spot elevations of finished grades are shown at the corners. Draw the building at a scale of 1 in. = 50 ft 0 in. and plot contours 85 and 86, the minimum slope away from and at right angles to the faces of the building being 2%.

PROBLEM 9.5.B.

Figure 9.12b shows a building 120 ft long and 80 ft wide with spot elevations of finished grades at the corners as indicated. Draw the building at a scale of 1 in. = 40 ft 0 in. and plot contours 135, 136, and 137; the minimum slope of the ground away from and at right angles to the faces of the building will be 2%.

Sidewalk Grades Shown by Contours

It is essential that sidewalks and paths have a cross slope to permit rain water to drain off. The cross slope is the slope taken at right angles to the longitudinal boundary of the sidewalk. Local highway departments frequently specify that the cross slope must have minimum and maximum cross slopes of 2% and

3%, respectively. City plans generally show the established curb elevations, and these elevations are the starting points for computing the grades of the sidewalks.

Figures 9.13a and b show to conditions of sidewalks adjacent to curbs. The curb elevations are given as data and are shown as spot elevations by small cross lines. For each condition shown, the length of the sidewalk is 115 ft and the width is 15 ft.

Figure 9.13a shows a sidewalk with a longitudinal 5% slope and a constant cross slope of 3% throughout its entire length. Beginning with the established curb elevations at points A and B, the elevations at the upper boundaries are computed; these are shown at points A' and B'. As an example, if the elevation of point A is 89.40, the width of the sidewalk is 15 ft, and the cross slope is 0.03, $D = G \times L$ or $D = 0.03 \times 15 = 0.45$ ft. Hence, the elevation at point A' is 89.40 + 0.45, or 89.85. The points of *even contours* (84, 85, 86, etc., with no decimals) are located on the two boundaries of the sidewalk, and the contour lines are drawn connecting them. We know that two parallel lines determine a plane, and, since the longitudinal slopes at both boundaries are the same, the sidewalk is a plane surface and the contour lines are parallel. The rain water will drain from the sidewalk in a direction perpendicular to the contour lines. This is shown by the small arrow.

Figure 9.13b shows a similar condition with the exception that the cross slope is 0.02 at one end and 0.03 at the other. This results in the two longitudinal boundaries not being

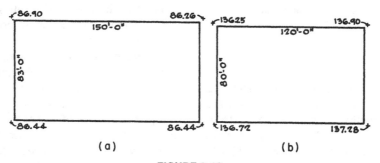

(a)   (b)

FIGURE 9.12.

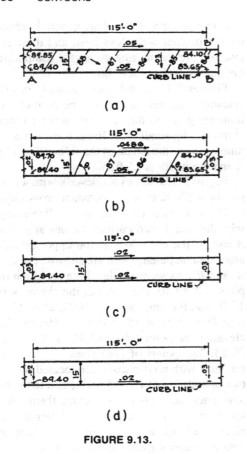

(a)

(b)

(c)

(d)

FIGURE 9.13.

parallel in space; the surface of the sidewalk is a warped surface, and the contour lines are not parallel. Note that angle θ_1 is greater than θ_2; hence, *smaller cross slopes result in greater angles made by the contour lines with the curb line*. If there were no cross slope the contours would be perpendicular to the curb, an undesirable condition, for this would result in no cross drainage of surface water.

PROBLEMS 9.5.C AND D.

Figures 9.13*c* and *d* represent sidewalks whose lengths between spot elevations are 115 ft and whose widths to the curb line are 15 ft. For the slopes and curb elevations shown, compute the elevations at the upper and lower boundaries and draw the 88 and 89 contour lines.

9.6 CONTOURS DRAWN FROM SPOT ELEVATIONS

For plane areas, it is possible to construct the contour lines when only spot elevations are known. When only spot elevations are shown on a drawing, it is assumed that the lines connecting them are of uniform slope. Spot elevations always take precedence over elevations determined by contours.

Figure 9.14 shows a four-sided plot with spot elevations given for points *A*, *B*, *C*, and *D*. Consider first the triangular area *ABC*. We know that *three points determine a plane* and that *contour lines are parallel on a plane surface*. Hence, if we establish one contour line on the area, the remaining contour lines may be drawn parallel to it. Suppose we are required to plot the contour lines on area *ABC*. Beginning with the side *AC*, we compute the gradient (0.0796) and the position of the points of even contour. Next, the gradient on line *AB* is computed and *one* point of even contour is located, say point *b*, contour 120. Now, by drawing a line from point *b* to point *a* (contour point 120 on line *AC*) we establish contour line 120. Since we know that all contour lines on this plane surface *ABC* are parallel, we draw lines from the points of even contour on line *AC* parallel to contour 120, thus establishing all the contour lines on area *ABC*.

The four-sided plot shown in Figure 9.14 is composed of two triangular areas, *ABC* and *BCD*; the contour lines on *ABC* have been established. To find the contours on the area *BCD*, begin by computing the gradient (0.0495) on line *BD* and locating the points of even contour. From contour point *c* on line *BD* draw a line to point *d* on line *BC*. This establishes contour 117. Through the points of even contour on line *BD* we now draw lines parallel to contour 117 and thus determine all the contour lines on area *BCD*.

A plot consisting of two or more triangular plane areas, as shown in Figure 9.14, would exist only on paved surfaces. On a natural-earth or lawn surface erosion would not permit sharp angles, such as shown by the con-

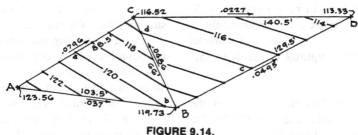

FIGURE 9.14.

tours at line *BC*, to remain. The contours at such a line should be rounded.

9.7 CONTOURS ON FILL

In order to obtain the desired grading adjacent to terraces or other features of a building, it is often necessary to alter the original conformation of the terrain by cuts or fill. The extent of the work must be clearly indicated by the contour lines. Different procedures may be employed, but the following example illustrates an approved method.

Example 1: The original ground area on which a terrace is to be constructed is a plane sloping surface, as indicated by the evenly spaced parallel contours in Figure 9.15, the contour interval being 1 ft 0 in. The terrace, *ABCD*, is 25 ft × 40 ft in plan; it begins on line *AD*, at which the original earth surface is 290.20, and has a downward pitch, toward edge *BC*, of 2% for drainage. The finished grading adjacent and at right angles to the

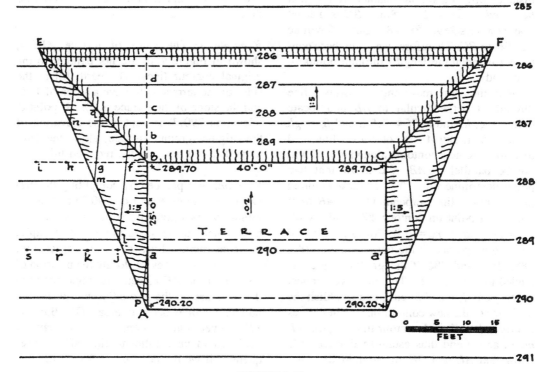

FIGURE 9.15.

sides of the terrace will have slopes of $1:5$, that is, 1 ft fall for each 5 ft horizontally, a grade of 0.2. Show, by means of contour lines, the grading required to result in this condition.

Solution: Since, by data, the elevations of the terrace at points A and D are each 290.20, the pitch is 2% and the width of the terrace is 25 ft; $D = G \times L$, or $D = 0.02 \times 25$, or 0.50 ft. Then the elevations at points B and C will each be $290.20 - 0.50$, or 289.70. Thus, it is seen that the only contour to cross the terrace area will be 290, and points a and a' will be $D \div G$, or $0.20 \div 0.02 = 10$ ft from the AD side of the terrace. Contour 290 will be parallel to AD.

Next, we draw a line through point B perpendicular to BC. On this line we plot points on contours 289, 288, 287, and 286 of the finished grading. These points are b, c, d, and e, respectively. To find the distance Bb, $D = 289.70 - 289.00$, or 0.70 ft. Then $L = D \div G$, or $L = 0.7 \div 0.2$ and $L = 3.5$ ft, the distance Bb. In a similar manner we find that $bc = cd = de = 5$ ft. Since BC is a level line, contours 289, 288, 287, and 286 will be parallel to BC. Therefore, we draw these contours (of indefinite length) through points b, c, d, and e.

With the same reasoning, we draw a line through B perpendicular to AB and locate points f, g, h, and i. $Bf = 3.5$ ft and $fg = gh = hi = 5$ ft. But AB is not a level line, and therefore the new contours along side AB will not be parallel to AB. We know that two points determine a line. To find these required points draw a line perpendicular to AB from point a (a point on contour 290) and lay off $aj = jk = kr = rs = 5$ ft. Points j, k, r, and s would be points on the new contours, 289, 288, 287, and 286 *if* the grading were extended to these points. Therefore we connect points f and j, g and k, etc., and establish the positions of the new contour lines. These lines intersect the original contour lines at points l, m, n, and o and thus establish line AE. We know that AE and BE are straight lines be-

cause the intersection of two planes is a straight line. Observe that contour 289 (line *jf*) intersects the original contour 289 at point l, a point common to both the original and the new contour. Such a point is called a *point of no cut* or *fill*; it marks a point on the edge of surface ABE. This is true also of points m, n, o, and p. A line joining these points defines the intersection of the side slope with the original surface of the ground. Line *no* is extended until it meets line BE and establishes the lower edge of the forward slope. Note that line BE bisects angle cqg.

The contour lines on the right-hand side of the terrace are found in a similar manner. *Hachures*, short free-hand lines, are usually drawn to the new contours and the undisturbed portions of the original contours are indicated by light solid lines. Those portions of the original contours effected by the grading are indicated by dash lines. In Figure 9.15, as well as in the following figures, lines and points used to determine the positions of new contours are shown for the purpose of construction and explanation; they should not be shown on the final drawings.

Example 2: The sides of terrace $ABCD$, shown in Figure 9.16 are not parallel to the original contour lines. The earth fill at the sides of the terrace is to have a slope of $1:5$, and the slope of the terrace is 2%. If point D (elevation 290.7) is kept at the elevation of the original ground surface, show the contours that will result in the required fill.

Solution: We proceed by computing the elevations of points B and C (290.2) and determining the positions of the new contour lines by the method explained in the last example; the construction lines show the procedure. This, however, does not locate the position of point E. Line AE bisects the angle between the contours on the two adjacent slopes, but no contour lines occur in area ADE. But line AD is level, and any contours in the plane of ADE would be parallel to line AD. Consequently, if we bisect angle daD with the line

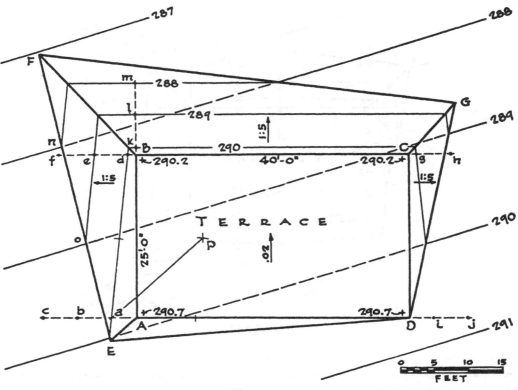

FIGURE 9.16.

ap we determine the *direction* of *AE*. Therefore, we draw a line through *A* parallel to *ap* and extend line *no*; their intersection determines point *E*. In Figures 9.16 to 9.23 inclusive, the hachures have been omitted for the sake of legibility.

Example 3: The terrace *ABCD*, shown in Figure 9.17, is 20 ft × 30 ft, and line *AD* is level (elevation 290.20). The slope at the center line of the terrace is to be 3%, and, in addition, there is a cross slope of 1% both ways on line *BC*. The purpose of this is to reduced the tendency of surface drainage to erode the top of the fill along line *BC*. The earth fill at the sides of the terrace is to have a slope of 1:3. Determine the positions of the contours to provide the required fill.

Solution: This problem differs from the former examples in that the contours of the orig-

inal earth surface are not parallel lines, the ground is not a plane surface. In this example, the intersections of the fill with the original surface will not be straight lines. The first step is to compute the elevation of points *B* and *C*, 289.45. After this, the new contours are found by the methods previously described. The points at which the new contours meet the original contours are connected by lines that mark the boundary of the fill; in this instance they are irregular lines.

9.8 CONTOURS ON CUTS

On the inspection of a drawing in which the contour lines indicate new grading, it may not at once be evident whether the work involved is a *cut* or *fill*. The following method will be found to be helpful in their identification.

Turn the drawing so that you are appar-

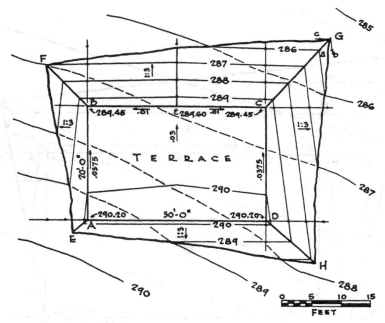

FIGURE 9.17.

ently looking *downhill*. This is indicated by the elevations of the original contour lines.

If the new contours have been moved toward you, from their original positions, the area is a cut. If the new contours have been moved away from you, the area is a fill. This system of identification may be verified by re-

ferring to the figures discussed in Section 9.7; they are all examples of fill.

Example 1: The 20 ft × 40 ft terrace *ABCD*, shown in Figure 9.18, is to have a 2% gradient toward edge *BC*. Line *BC*, elevation 211.5, coincides with the natural grade. The

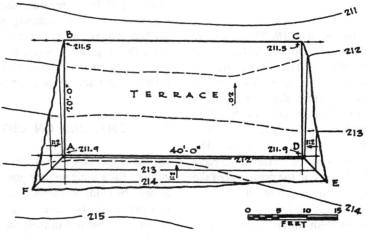

FIGURE 9.18.

side slopes of the cut will have a gradient of 1:2 *toward* the edges of the terrace. Determine the contours to show the shape and extent of the cut.

Solution: The procedure explained in Section 9.7 is followed, and the construction lines and contours are shown in Figure 9.18. This type of cut has the disadvantage of permitting surface water to drain over the terrace. To prevent this, the water may be intercepted by a depression (gutter) at the bottom of the cut. This is illustrated in the following example.

Example 2: Figure 9.19 shows a 20 ft × 40 ft terrace, *ABCD*, with a 3% gradient down toward edge *BC* (elevation 211.10), which is level. A V-type gutter is to be constructed at three sides of the terrace. The high point of the gutter, *F*, is 1.5 ft below *E*, a point on the center line of the terrace. The *bottom* of the gutter is to have a 2% slope, and the side slopes of the gutter are to be 1:2. Determine the positions of the contours to provide this condition.

Solution: The elevation of line *AD* is computed to be 211.70, and the elevation of point *F* will be 211.70 − 1.50, or 210.20. Distance *EF* will be 1.5 ÷ 0.5, or 3 ft. Point *G*, the bottom of the gutter opposite *D*, scales approximately 20 ft from *F*; hence, the elevation of *G* will be 210.20 − (20 × 0.02), or 209.80. Distance *DG* will be 1.90 ÷ 0.5, or 3.8 ft. Point *G* is now plotted, and a line from *F* to *G* locates the bottom line of the gutter. Contours 210 and 211 may not be plotted on the terrace side of the slope to the gutter area *EFHICD*, and line *DH* is thus established. By scaling, the distance *FH* is approximately 24.5 ft; hence, the elevation of point *H* is 210.20 − (24.5 × 0.02), or 209.71. Distance *HJ* also scales 24.5 ft; therefore, the elevation of point *J* becomes 209.71 − (24.5 × 0.02), or 209.22. Distance *CJ* is (211.10 − 209.22) ÷ 0.5, or 3.76 ft. Extended line *HJ* dies out on the natural grade at point *I*. Point *I* may generally be determined with sufficient accuracy

by inspection. By the methods previously explained, the contours on area *FHIKL* are now plotted. The left-hand side of the terrace is plotted in a similar manner.

9.9 CONTOURS FOR ROADS

When contours cross a road they repeat the shape of the cross section at what appears to be an exaggerated scale; Figure 9.20*a* shows such a condition. Figure 9.20*b* is a cross section of the road showing a paved sidewalk on one side with a shoulder and V-gutter on the other. The crown of the road is 5 in. higher than the sides, and the cross section is the curve of a parabola. Note the gradients shown on the drawings. Point *A*, on the center line of the road, has an elevation of 592.62, and the gradient of the road is 0.028.

By the procedure explained in Section 9.5, we begin by plotting the points of even contour, 592, 591, 590, and 589 (points *a*, *b*, *c*, and *d*) on the center line of the road. Each of these points will be the apex of a parabola. Let us consider contour 591; point *b* is on this contour. The crown of the road is 5 in., or 0.42 ft above the bottom of the curve. Then distance *be* will be *D* ÷ *G*, 0.42 ÷ 0.028, or 15 ft, and point *e* is thus plotted on the center line of the roadway. A line drawn through *e* at right angles to the center of the road establishes points *f* and *g* at the base of the parabola, and this permits us to draw curve *fbg*, which is contour 591 on the roadway. (The method of drawing a parabola is explained in Section 10.1.) The height of the curb is 0.50 ft; then, to find point *h*, distance *fh* = *D* ÷ *G* = 0.50 ÷ 0.028 = 18 ft. The sidewalk has a cross slope of 2%; hence the other line of the sidewalk is *G* × *L* = 0.02 × 10, or 0.20 ft above the curb line. Then *hi* = *D* ÷ *G* = 0.20 ÷ 0.028, or 7.2 ft. A line through *i* perpendicular to the curb line establishes point *j*, and line *hj* is contour 591 on the sidewalk. Point *l*, the bottom of the gutter, is located similarly. Distance *gk* = *D* ÷ *G* = 1.25 ÷ 0.028 = 44.5 ft, and point *l* is directly op-

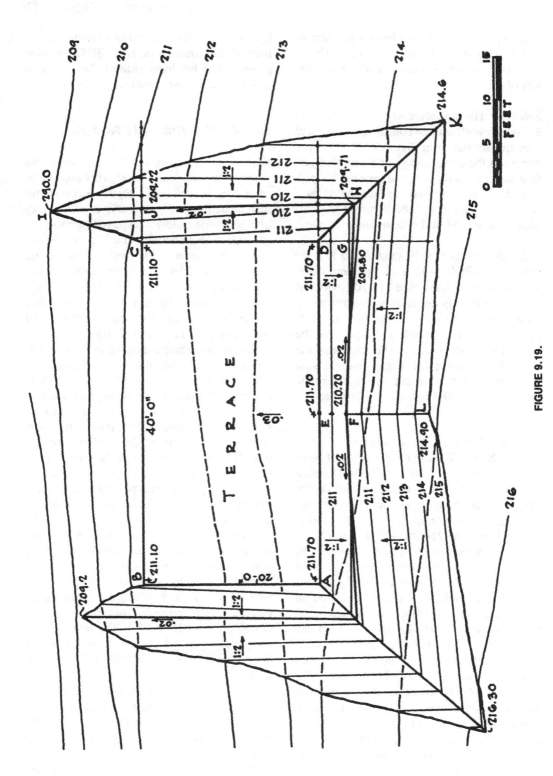

FIGURE 9.19.

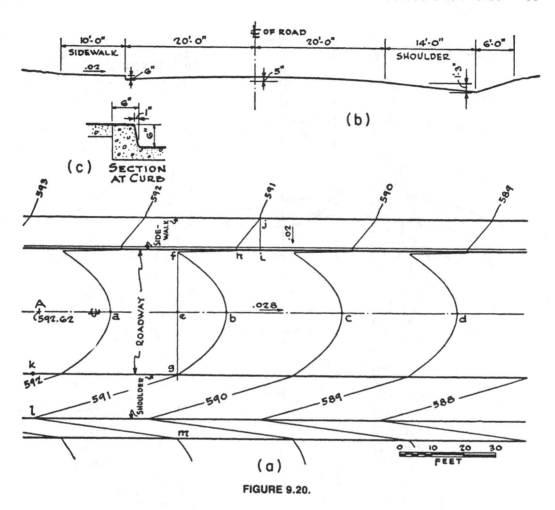

FIGURE 9.20.

posite *k* at the low point of the gutter. Point *m* is opposite *g*. This completes contour 591; the remaining contours are found in the same manner.

Cut and Fill at the Sides of a Road

Figure 9.21 shows a straight roadway crossing contours that necessitate both a cut and fill. The road has a gradient of 0.04, and the side slopes of both the cut and fill, taken at right angles to the center line of the roadway, will be 1:4. Point *A* has an elevation of 352.80, and the dimensions of the roadway and shoulders are indicated on the section.

The first step is to locate the points of even contours on the center line of the road and to plot the even contours on the road and shoulders, as explained previously in this section.

At any point of even contour at the edge of the shoulder, for example, point *a* (elevation 352), draw a line perpendicular to the center line of the road. On this line lay off distances *ab* = *bc* = *cd*, etc. = 4 ft (the slope is 1:4). At this slope point *b* will be 1 ft below point *a* and would be a point on contour 351 if the contour extended this far. Now point *h* at the edge of the shoulder is a point on contour 351; hence, by joining points *h* and *b* we establish contour 351 on the slope. We note that this line intersects the natural contour 351 at point *g*. This is a point of no cut or fill, and it marks

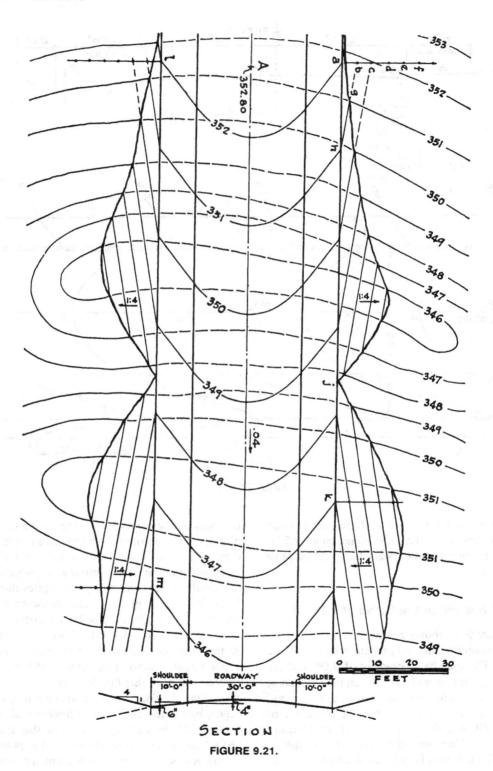

SECTION

FIGURE 9.21.

curved line. The other contours are found in the same manner. The intersections of the contours with the original slope of the ground establish the boundary lines of the fill.

9.10 CONTOURS AT SITE CONSTRUCTION

Interruption of the smooth form of ground surface contours occurs in various forms of construction. Previous examples in this chapter have illustrated the situations that occur at portions of cuts and fills, roadways, and sidewalks. This section presents the conditions that occur at some additional forms of site construction.

Contours at Steps

When steps occur in a path or walk, either a bank of earth or a retaining wall is required to accommodate the abrupt change in level of the adjoining earth surfaces.

Figure 9.23a shows a 10 ft wide path and steps, a section through which is shown in Figure 9.23b. The banks at the sides of the steps are to have a slope of 1:3. On the right-hand side of the steps a line of indefinite length is drawn at right angles to the direction of the path. This line will be the top of the bank; it will have a 1% slope. Point A has an elevation of 49.12, and point a (elevation 49) is plotted. It is seen that the top of the bank will meet the natural slope of the ground at some point whose elevation is between 48 and 49. Point c (elevation 48.5), 62 ft from A, is plotted, but from the positions of the original contours it is apparent that the top line of the bank will not extend this far. Point b (elevation 48.6), 52 ft from A, is next plotted, and we see that the top line of the slope meets the natural slope at point B whose approximate elevation is 48.55. Distance AB = 0.57 ÷ 0.01, or 57 ft. Points of even contours are located on the cheek wall of the steps, and the contours on the bank are drawn to intersect the natural contours at the bottom of the slope. Contour 49 runs out at the top of the bank at point a, which is above the natural grade at

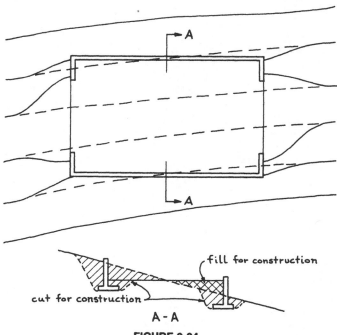

A - A

FIGURE 9.24.

this point. This requires that earth be filled in here, and a portion of contour 49, *ad*, is drawn so that surface water will be diverted from the path. Contour 50 is altered similarly.

If the above method is followed on the left-hand bank, the bank would be extremely long. To avoid this, the bank is bent; it forms a warped surface, and the contours on it will be curved lines. Suppose we limit the length of *DE* to 57 ft. The elevation at *D* will be 49.12 − 57 × 0.01), or 48.55. Then point *E* is interpolated between contours 48 and 49, and line *DE*, 57 ft in length, is drawn in. Contours 47 and 48 on the bank are drawn by the construction indicated on the drawing.

Contours at Retaining Walls

Changes in ground surface forms may be achieved with sloped soils, but are also fre-
quently achieved with some form of construction. When a change in elevation must be abruptly achieved, a retaining structure is used. This may consist of a simple curb for small differences in elevation, but usually consists of a cantilever retaining wall for changes over 2 ft or so.

Figure 9.24 shows the use of two retaining walls to assist in developing a flat portion of ground in a sloping hillside condition. The uphill retaining wall helps to achieve the cut into the hill, and the downhill wall retains the fill used to build up the site.

Walls could be used on all four sides of the site in Figure 9.24, but the plan shows the use of simple banking of the fill at the ends where the recontouring is minor in vertical dimension of change. The turned ends of the walls help to ease the transition at the corners of the site.

10

Vertical Curves

10.1 PARABOLIC CURVES

When the slope of a roadway changes and the difference between the grades exceeds 1% ($\frac{1}{2}$% on important and high-speed roads), the abrupt change is eased by the use of a vertical (profile) curve. Vertical curves are used when a downward slope changes to an upward slope (a sag curve) or when an upward slope changes to a downward slope (a peak curve). Owing to the fact that they are readily computed and plotted, the vertical curves used for roadways are invariably parabolas.

The curve shown in Figure 10.1 is a *parabola*. It is tangent to line *AB* at point *A*. The vertical height is *h*, and *D* is the half-width of the parabola. *P* is any point on the curve; its horizontal distance from the point of tangency is *x*, and *y* is its vertical distance from the tangent line. For any distance *x*, the distance *y* may be computed by the formula

$$y = \left(\frac{x}{D}\right)^2 h$$

which, in effect, states that *the vertical offsets are proportional to the squares of their distances from the point of tangency.* In using this formula, *be certain that all the terms are of the same units, feet or inches.*

Example 1. In the parabola shown in Figure 10.1, what is the offset *y* if $D = 20$ ft, $h = 8$ in., and $x = 13$ ft?

Solution: Since $h = 8$ in., $h = (\frac{8}{12})$ ft, or 0.67 ft. Then, substituting in the formula,

$$y = \left(\frac{x}{D}\right)^2 h, \quad y = \left(\frac{13}{20}\right)^2 \times 0.67$$

and

$$y = 0.28 \text{ ft or } 3\tfrac{3}{8} \text{ in.}$$

Example 2. In the parabola shown in Figure 10.2, *h* is the height of the parabola and *D* is the half-width. If *D* is divided into ten equal parts, compute the values of the *y* distances in terms of *h*.

101

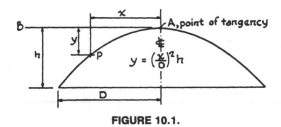

FIGURE 10.1.

Solution: Beginning at the center line of the curve, the successive values of x are $D/10$, $2D/10$, $3D/10$, etc. Let the y distances corresponding to the various x distances be y_1, y_2, y_3, etc. Then, substituting in the formula for the parabola,

$$y_1 = \left(\frac{D/10}{D}\right)^2 \times h = \left(\frac{1}{10}\right)^2 \times h$$

$$= \frac{1}{100} h$$

$$y_2 = \left(\frac{2D/10}{D}\right)^2 \times h$$

$$= \left(\frac{2}{10}\right)^2 \times h = \frac{4}{100} h$$

$$y_3 = \left(\frac{3D/10}{D}\right)^2 \times h$$

$$= \left(\frac{3}{10}\right)^2 \times h = \frac{9}{100} h$$

etc.

The remaining values of y are shown in Figure 10.2. This figure shows the y distances when D is divided into ten equal parts. It will be found to be of great assistance in plotting any parabolic curve. Eleven points on the curve are shown, but usually a smaller number is sufficient to lay out the curve.

Example 3. A roadway is 20 ft in width, the profile taken on a section *across* the road is a parabolic curve, and its height at the center line is 6 in. It is desired to lay out a template for this cross section. Compute the positions of various points on the curve.

Solution: Since the road is 20 ft in width, a horizontal line is drawn 10 ft in length, corresponding to D in Figure 10.2. On this line points 1 ft 0 in. apart are laid off. Then the y distances for these points are computed by using the values of y shown in Figure 10.2. Note that $h = 6$ in., or 0.5 ft. Then $y_1 = \frac{1}{100} \times 0.5 = 0.005$ ft, $y_2 = \frac{4}{100} \times 0.5 = 0.02$ ft, and $y_3 = \frac{9}{100} \times 0.5 = 0.045$ ft. Similarly, $y_4 = 0.08$ ft, $y_5 = 0.125$ ft, $y_6 = 0.18$ ft, $y_7 = 0.245$ ft, $y_8 = 0.32$ ft, $y_9 = 0.405$ ft, and $y_{10} = 0.5$ ft.

10.2 PLOTTING VERTICAL CURVES IN A ROADWAY

A downward slope of a roadway has a grade of 0.03; it intersects an upward slope having

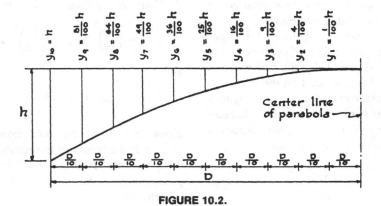

FIGURE 10.2.

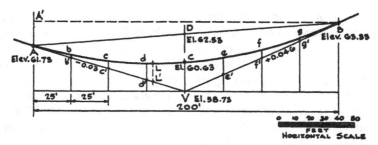

FIGURE 10.3.

a grade of 0.046 at point V, as shown in Figure 10.3. When traversing a road, a downward slope is marked minus (-0.03) and an upward slope is marked plus ($+0.046$). The point of intersection of the two grades (P.I.) is point V; its elevation is 58.73. A vertical sag curve, 200 ft long in horizontal projection, is to be placed in the road from points A to B. Suppose we are required to determine the elevations of points on a parabolic curve, joining points A and B, at 25 ft (horizontal) intervals.

All measurements of length on a plan are horizontal projections. Thus, in Figure 10.3 the lengths of ADB, ACB, and AVB when shown in plan are all equal to the horizontal projection $A'B$. In a vertical curve, $AV = VB$ in horizontal projection and *point C is always midway between points D and V*. The elevation of point V is given to be 58.73. The grades of AV and VB are known, and thus we can compute the elevations of points A, B, C, D, b', c', d', e', f', and g'; the computations are shown in Figure 10.4. The elevation of point D is midway between the elevations of points A and B, and point C is midway between D and V. Lines AV and VB are tangent to the parabola at points A and B; conse-

$$
\begin{aligned}
\text{ELEV. } @A &= 58.73 + (100 \times .03) &&= 58.73 + 3.00 = 61.73 \\
@b' &= 58.73 + (75 \times .03) &&= 58.73 + 2.25 = 60.98 \\
@c' &= 58.73 + (50 \times .03) &&= 58.73 + 1.50 = 60.23 \\
@d' &= 58.73 + (25 \times .03) &&= 58.73 + 0.75 = 59.48 \\
@V &= 58.73 \text{ (given)} \\
@e' &= 58.73 + (25 \times .046) &&= 58.73 + 1.15 = 59.88 \\
@f' &= 58.73 + (50 \times .046) &&= 58.73 + 2.30 = 61.03 \\
@g' &= 58.73 + (75 \times .046) &&= 58.73 + 3.45 = 62.18 \\
@B &= 58.73 + (100 \times .046) &&= 58.73 + 4.60 = 63.33
\end{aligned}
$$

$$@D = \frac{61.73 + 63.33}{2} = 62.53 \quad \text{(Point } D \text{ is midway between } A \text{ and } B\text{)}$$

$$@C = \frac{62.53 + 58.73}{2} = 60.63 \quad \text{(Point } C \text{ is midway between } D \text{ and } V\text{)}$$

$$VC = 60.63 - 58.73 = 1.90$$

$$b'b = gg' = \left(\tfrac{1}{4}\right)^2 \times 1.90 = \tfrac{1}{16} \times 1.90 = 0.12$$

$$c'c = ff' = \left(\tfrac{1}{2}\right)^2 \times 1.90 = \tfrac{1}{4} \times 1.90 = 0.48$$

$$d'd = ee' = \left(\tfrac{3}{4}\right)^2 \times 1.90 = \tfrac{9}{16} \times 1.90 = 1.07$$

$$
\begin{aligned}
\text{ELEV. } @b &= 60.98 + 0.12 = 61.10 \\
@c &= 60.23 + 0.48 = 60.71 \\
@d &= 59.48 + 1.07 = 60.55 \\
@e &= 59.88 + 1.07 = 60.95 \\
@f &= 61.03 + 0.48 = 61.51 \\
@g &= 62.18 + 0.12 = 62.30
\end{aligned}
$$

FIGURE 10.4. Computation of elevations.

quently the offsets, $b'b$, $c'c$, $d'd$, etc. (the y distances), are proportional to the squares of their distances from point A, the point of tangency. Point b' is $\frac{25}{100}$ or $\frac{1}{4}$ of the distance from point A to point V, and $VC = 1.9$ ft (Figure 10.4). Then $b'b = (\frac{1}{4})^2 \times 1.9 = 0.12$ ft. The elevation of b' is 60.98; hence $60.98 + 0.12 = 61.10$, the elevation of point b. The elevations of other points on the curve are found in a similar manner and are shown in Figure 10.4.

Peak Curves

When a roadway goes over the crest of a hill the vertical curve is inverted from the curve shown in Figure 10.3. To find the elevations of various points on the curve, the elevations of points on the grade lines AV and VB (Figure 10.5) are found and the offsets are then *subtracted*, the procedure being similar to that described in Section 10.2.

10.3 HIGH AND LOW POINTS ON VERTICAL CURVES

When a drain or catch basin is to be placed at the *low point* of a curve on a roadway, it becomes necessary to locate the exact position of the low point. The horizontal distance from the point at which a curve begins to the low (or high) point of the curve is

$$\frac{lg_1}{g_1 - g_2}$$

in which l is the horizontal projection of the total length of the curve and g_1 and g_2 are the grade percentages at the beginning and end of the curve, respectively. In using this formula, refer to Section 10.2 concerning plus and minus grades.

In Figure 10.3 point C would be the low point of the curve only when AV and VB were of equal slopes. In this instance the low point is obviously to the left of point C, let it be point L. Substituting in the formula,

horizontal projection of distance AL

$$= \frac{200 \times (-0.03)}{(-0.03) - (+0.046)}$$

$$= \frac{200 \times (-0.03)}{-0.076} = 79 \text{ ft}$$

To find the elevation of point L, first find the elevation of point L'. Then, since the elevation of A is 61.73,

$$61.73 - (0.03 \times 79)$$
$$= 59.36 \quad \text{the elevation of } L'$$

To find distance $L'L$ we use the principle given in Section 10.1 and distance $L'L = (79/100)^2 \times 1.9 = 1.19$ ft. Therefore, $59.36 + 1.19 = 60.55$, the elevation of point L, the low point of the curve.

10.4 PROFILES

With respect to surveying, a profile is the vertical projection of the intersection of a vertical plane with the surface of the earth. In plotting contours which cross highways, profiles are used advantageously. Since the slopes of roads are relatively small in comparison with their lengths, the vertical scale of a profile is generally exaggerated, being made five or ten times larger than the horizontal scale. Profiles are useful in checking vision or sight distances at the crest of peak curves. For highway work this is a necessity, and it is also advisable for private roads and drives. Specially ruled paper for this type of work is available.

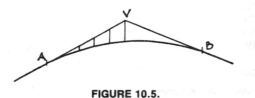

FIGURE 10.5.

10.5 PLOTTING CONTOURS BY USE OF A PROFILE

To explain the method of plotting contours crossing a roadway in which there is a vertical curve, consider the 200 ft curve shown in Figure 10.3. A cross section through the roadway and shoulders is shown in Figure 10.6*a*. Note that the road has a width of 40 ft and that 10 ft is the width of the shoulders. For drawing the profile, a larger scale has been used for the vertical dimensions; the heights are shown on Figure 10.6*c*. The roadway with the shoulders is shown in plan in Figure 10.6*b*.

The vertical curve shown in Figure 10.3 is the curve at the center line of the roadway. This curve, *ALB*, is shown again *in profile* in Figure 10.6*c* by the solid curved line. The elevations of points *A*, *B*, and *L* are known, and the horizontal lines numbered 58 to 65 represents the *levels* of the various contour elevations. These contour levels are 1 ft apart and are drawn at a larger scale as an aid in plotting the contours. In Figure 10.6*a* we see that the sides of the road are 5 in., or 0.42 ft, below the center line of the road and that outer edges of the shoulders are 13 in., or 1.08 ft, below the center line. Now draw a vertical line through any point, such as *a*, on the curve shown in Figure 10.6*c*, and lay off *ab* = 0.42 ft and *ac* = 1.08 ft. Thus, points *b* and *c* are points on the curves of the shoulders of the road. A sufficient number of such points are plotted to enable us to draw the two curved dash lines which represent the shoulder lines in profile.

In Figure 10.6*c* it is seen that the horizontal lines that represent the levels of the contours intersect the center line of the roadway and edges of the shoulders at various points. As an example, contour 61 intersects the center line of the roadway (to the left of the low point *L*) at point *d* and the shoulder lines at points *e* and *f*. By projecting these points up to the plan, (see Figure 10.6*b*), we establish points *d'*, *e'*, *e''*, *f'*, and *f''*; these are points

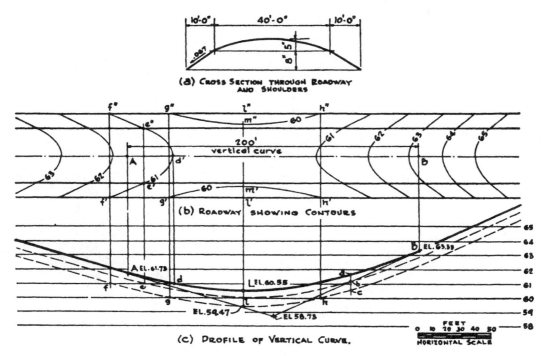

(a) CROSS SECTION THROUGH ROADWAY AND SHOULDERS

(b) ROADWAY SHOWING CONTOURS

(c) PROFILE OF VERTICAL CURVE.

FIGURE 10.6.

on contour line 61, and a line connecting them is contour 61. The remaining contours are drawn in a similar manner.

Note that contour level 60 intersects the outer edge of the shoulder at points g and h. In Section 10.3 we found the elevation of L, the low point on the center line of the road, to be 60.55. Hence the elevation of point l is $60.55 - 1.08$, or 59.47. Now points g, l, and h are projected up to the plan and we establish points g', l', and h'. The slope of the shoulders is 8 in. in 10 ft, a grade of $0.067/10$, or 0.067. Contour 60 is $(60 - 59.47)$, or 0.53

ft above point l. Therefore $l'm'$ and $l''m''$ are each $0.53/0.067$, or 7.9 ft. Curves drawn through $g'm'h'$ and $g''m''h''$ give us contour 60. The contours crossing the vertical curve, shown in Figure 10.6b, are not necessarily parabolas or straight lines. The method described is the usual procedure for plotting contours on vertical curves. If greater accuracy is desired, additional points on the contours are found by selecting other lines, such as the quarter points of the road and half points on the shoulders, with the corresponding profile curves.

11

COMPUTATIONS FOR CUT AND FILL

11.1 CUT AND FILL

An ideal grading situation is that in which the volume of cut equals the volume of fill. On the completion of such an operation there remains no excess earth to be transported elsewhere and no additional material need be brought to the site. Such a condition can only be obtained by computing the volume of cut and fill before the work is begun.

11.2 EXCAVATIONS FOR BUILDINGS

When the excavation for a building is comparatively small and the surface of the ground has a uniform slope, the excavation may be considered to be a prism. By multiplying the area of the excavation by the *average* height of the corners, we obtain the approximate volume of earth to be excavated.

Example 1. A 12 ft × 20 ft rectangular excavation, *ABCD*, is shown in Figure 11.1. The contours show the natural surface of the ground to have an approximately uniform slope. Compute the approximate volume of earth to be excavated if the bottom of the excavation has an elevation of 92.5.

Solution: By interpolating between the contours, the elevations at the corners are:

$$A = 100.7 \quad B = 102.3 \quad C = 104.2$$
$$D = 102.5$$

By subtracting the elevation of the bottom of the excavation, 92.5, from these elevations, the heights of the excavation at the corners are:

$$A = 8.2 \text{ ft} \quad B = 9.8 \text{ ft} \quad C = 11.7 \text{ ft}$$
$$D = 10.0 \text{ ft}$$

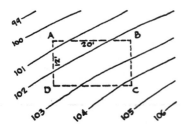

FIGURE 11.1.

Then the average height multiplied by the area of the excavation is

$$\frac{8.2 + 9.8 + 11.7 + 10.0}{4} \times 12 \times 20$$

$$= 2{,}382 \text{ ft}^3 \text{ or } \frac{2{,}382}{27}$$

$$= 88.2 \text{ yd}^3$$

When the excavations are more extensive, or when the surface of the ground is irregular, a somewhat similar system of computation is recommended. In this method the area is divided into a number of smaller areas, the average height is computed by taking the heights at a greater number of places, and the result-

ing volume is computed with greater accuracy.

Example 2. The irregular area *ABCDEF*, shown in Figure 11.2 represents the area of a proposed excavation of which the elevation of the bottom is 208.0. Compute the number of cubic yards of earth to be excavated.

Solution:

STEP 1.   Divide the area into a grid (a system of adjoining squares) consisting of squares 10 ft on each side. Squares of any convenient size may be used.

STEP 2.   Letter the corners of the squares so that the letter *a* designates the corners which occur in only one square, *b* designates the corners common to two squares, *c* designates the corners common to three squares, and *d* designates the corners common to four squares. See Figure 11.2

STEP 3.   By the method used in the last example compute the height of the excavation at all the corners and also compute the sum of the heights of all the *a*'s, *b*'s, *c*'s, and *d*'s. The Greek letter Σ, used in mathemat-

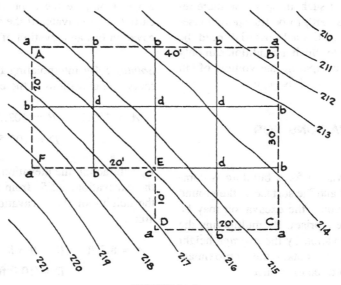

FIGURE 11.2.

ics, may be read "the sum of." Thus,

a's					
215.7	211.3	214.6	217.2	219.3	
208.0	208.0	208.0	208.0	208.0	
7.7	3.3	6.6	9.2	11.3	Σa's $= 38.1$

b's							
214.6	213.6	212.5	213.1	213.9	215.7	217.4	217.4
208.0	208.0	208.0	208.0	208.0	208.0	208.0	208.0
6.6	5.6	4.5	5.1	5.9	7.7	9.4	9.4

Σb's $= 54.2$

c's
215.9
208.0
7.9

Σb's $= 7.9$

d's			
215.9	214.7	213.8	214.6
208.0	208.0	208.0	208.0
7.9	6.7	5.8	6.6

Σd's $= 27.0$

STEP 4.   Compute the volume in cubic yards by use of the formula

$$\text{volume} = \frac{\text{area of 1 square}}{27} \times \frac{\Sigma a\text{'s} + 2\Sigma b\text{'s} + 3\Sigma c\text{'s} + 4\Sigma d\text{'s}}{4}$$

or,

$$\text{volume} = \frac{10 \times 10}{27} \times \frac{38.1 + (2 \times 54.2) + (3 \times 7.9) + (4 \times 27.0)}{4}$$

$$= 258$$

which is the number of cubic yards of earth to be excavated. On large or complicated areas this method will prove to be long and tedious; the procedure explained in Sections 11.5 and 11.6 will be found more efficient.

11.3 VOLUME OF A PRISMOID

A *prismoid* is a solid with parallel but unequal ends or bases whose other faces are quadrilaterals or triangles. In computing volumes of earth, various cross sections of cut or fill are considered; they indicate parallel areas at definite distances apart. The volumes bounded by these areas approximate prismoids and are so considered in the computations. Two methods are used for computing the volume of a prismoid.

(a) Average End-Area Method. Figure 11.3 shows a prismoid of given dimensions. The top and bottom faces, A_1 and A_2, are parallel planes. A formula commonly used for finding the volume of a prismoid is

$$\text{volume} = \frac{A_1 + A_2}{2} \times l$$

in which A_1 and A_2 are the areas of the two parallel faces and l is the normal distance between them, $(A_1 + A_2) \div 2$ being the *aver-*

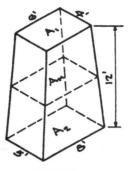

FIGURE 11.3.

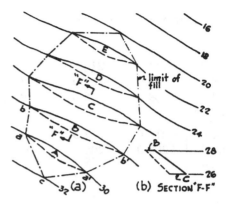

FIGURE 11.4.

age end area. The volumes obtained by use of this formula are not exact but generally are somewhat in excess of the exact values. Because of its simplicity, however, this formula is used extensively. Using it for the solid shown in Figure 11.3,

$$\text{volume} = \frac{(4 \times 6) + (5 \times 8)}{2} \times 12 = 384 \text{ ft}^3$$

(b) Prismoidal Formula. The formula that gives the exact volume of a prismoid is

$$\text{volume} = \frac{A_1 + 4Am + A_2}{6} \times l$$

in which Am is the area of the section midway between the two parallel faces, the remaining terms being similar to those in the previous formula. Thus, for the prismoid shown in Figure 11.3,

volume

$$= \frac{(4 \times 6) + (4 \times 4.5 \times 7) + (5 \times 8)}{6} \times 12 = 380 \text{ ft}^3$$

11.4 VOLUMES OF CUT OR FILL FOUND FROM CONTOURS

A simpler and more efficient method of computing volumes of cut and fill is to use the contours. In fact, this is one of the principal reasons for showing contours. Figure 11.4*a* shows a portion of a plot, the contours indicating a fill. The dotted lines show the original contours and the solid lines the contours after the fill has been placed. Points *a a'*, *b*, *b'*, etc., indicate the extremities of the fill. The areas between the original and finished contours are identified as *A*, *B*, *C*, *D*, and *E*. Figure 11.4*b* shows a section through the fill at *F–F*, the dotted line being the original slope of the ground. Note that *B* and *C* are parallel planes, parallel faces of a solid the height of which is the vertical distance between them, actually the contour interval. If we consider the curved contours as a series of short straight lines, the earth included in the solid is, approximately, a prismoid. Therefore, if we compute the areas of *B* and *C* we can apply the "average end-area method" to compute the volume of the solid. Point *c* on contour 32 is a "point of no cut or fill," and a section, similar to *F–F*, taken through point *c*, is a triangle, the solid in this portion of the fill being approximately pyramidal in shape, the base being area *A*. This same condition is found between contours 20 and 22. The volume of the pyramid is $\frac{1}{3}$ (base × height). Thus, the approximate volume of the entire fill may be found by adding together the volumes of all the individual prismoids and pyramids.

Let *A*, *B*, *C*, *D*, and *E* = the areas between the original and finished contours, in square feet; let *i* be the contour interval, in feet; and let *V* be the approximate total volume of the cut or fill, in cubic feet. Then,

$$V = \frac{Ai}{3} + \frac{A + B}{2}i$$
$$+ \frac{C + D}{2}i + \frac{D + E}{2}i + \frac{Ei}{3}$$

or

$$V = i\left(\frac{5A}{6} + B + C + D + \frac{5E}{6}\right)$$

From the preceding discussion it is seen that volumes computed by this method are only approximately correct, the contours themselves being only approximately exact. Thus, taking $\frac{5}{6}$ of the top and bottom areas (A and E) may be considered to be an unnecessary refinement. If these fractions are omitted, the rule that may be used to find the approximate volume is: *Add together all the areas, A, B, C, etc., and multiply their sum by the contour interval.* If the volume thus found is in units of cubic feet, divide by 27 to find the volume in cubic yards.

11.5 THE PLANIMETER

There are several methods of computing the contents of an area bounded by an irregular line. The method of dividing the area into trapezoids, explained in Section 6.7, can be used, or cross-section paper might be employed. But both of these methods are tedious, and the best procedure is to use an ingenious instrument called a *planimeter*, (see Figure 11.5.)

The planimeter is an instrument used for measuring areas of plane figures. It consists of a metal frame to which are attached a weighted *anchor point*, a *measuring wheel* with counter and vernier, and a *tracing point*. To measure an area, the anchor point is pressed into the drawing board *outside* the area. The tracing point is placed at some marked point on the boundary of the area, and a reading of the counter and vernier is taken. Next, the tracing point is traced around the outline of the area *in a clockwise direction*, returning *exactly* to the starting point. Another reading is now taken, and the difference in the readings gives the area of the figure *in square inches*. Careful use of the instrument will result in an error not exceeding 1%.

The final step is to convert the square inches to the scale of the drawing. As an example, suppose the area found is 2.56 in^2 and the scale of the drawing is 1 in. = 40 ft 0 in. Then, $2.56 \times 40^2 = 4{,}096$ ft^2, the area.

11.6 COMPUTING THE VOLUMES OF CUT AND FILL WITH THE PLANIMETER

Figure 11.6 shows a portion of a site plan with the contours indicated in the conventional

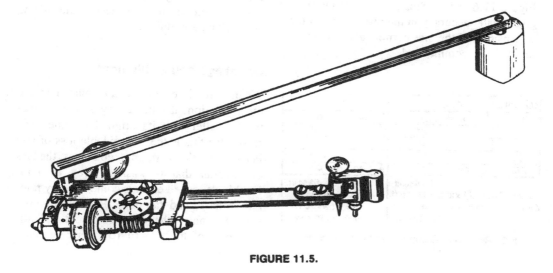

FIGURE 11.5.

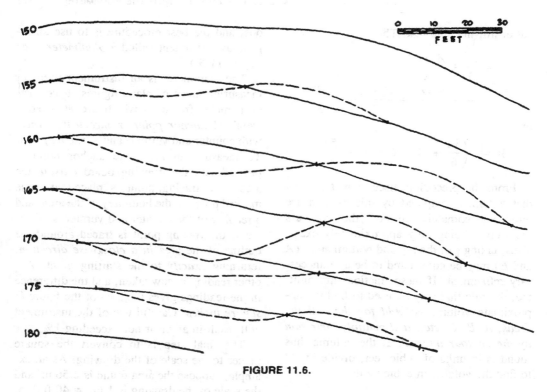

FIGURE 11.6.

manner. The left-hand portion of the required grading is a fill and the portion on the right is to be a cut. The areas of cut and fill are found by use of the planimeter and the results are tabulated in Figure 11.7. Note that this tabulation conforms with the formula given in Section 11.4. The contour interval shown in Figure 11.6 is 5 ft. Smaller contour intervals give greater accuracy in the results. When the dimensions of the site permit, a 1 ft contour interval is recommended.

CONTOUR	FILL	CUT
155	$\frac{5}{8} \times .13$ = .11	$\frac{5}{8} \times .08$ = .07
160	.27	.18
165	.47	.11
170	.21	.05
175	$\frac{5}{8} \times .09$ = .08	$\frac{5}{8} \times .08$ = .05
	1.14 sq. in.	.46 sq. in.

(SCALE 1"=40'-0") ×1600 = 1824 sq.ft.   ×1600 = 736 sq.ft.
(CONTOUR INTERVAL) ×5 = 9120 c.f.   ×5 = 3680 c.f.
÷27 = 338 c.y.   ÷27 = 136 c.y.

FIGURE 11.7. Computation for cut and fill.

Balancing Cut and Fill

The computations given in Figure 11.7 show that there is a greater volume of fill than cut. If fill cannot be obtained from other parts of the plot, the contours should be revised so that there are approximately equal volumes of cut and fill. This may require several trial locations of contours or even necessitate raising or lowering a building.

Shrinkage and Settlement

When earth is excavated and placed in another position, it will occupy a space somewhat smaller than its original volume. The reason for this is the unavoidable loss of soil in transportation and, on steep slopes, the loss of material that is washed away by rain. Freshly deposited earth is compacted by rainfall, rollers, and tamping. Because of this, when balancing the volumes of cut and fill, an excess of fill, 5% to 10%, is commonly provided; that is, 100 ft³ of cut and 107 ft³ of fill would about balance.

Building Excavations Determined by the Planimeter

When an excavation has vertical sides (the usual condition), its volume may be included, in computing the volume of cut and fill on a plot, by using the planimeter. *Contours that meet the vertical sides of an excavation lie in the vertical face of the excavation on the uphill side.* This is shown in Figure 9.5*b*.

Example. Figure 11.8*a* shows a 40 ft × 60 ft excavation, *ABCD*, the elevation of the bottom of the excavation being 80.5. As shown by the contours, the grading work to be per-

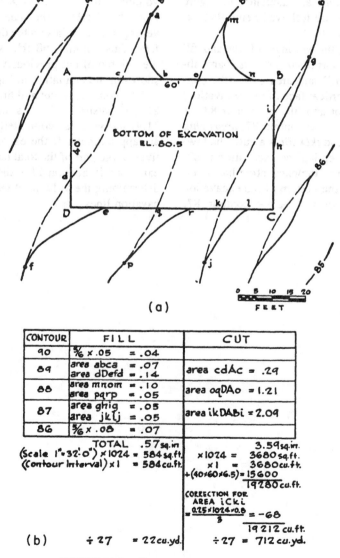

(a)

CONTOUR	FILL		CUT
90	⅝ × .05	= .04	
89	area abca	= .07	area cdAc = .29
	area dDefd	= .14	
88	area mnom	= .10	area oqDAo = 1.21
	area pqrp	= .05	
87	area ghig	= .05	area ikDABi = 2.09
	area jklj	= .05	
86	⅝ × .08	= .07	

TOTAL .57 sq.in

(Scale 1"=32'-0") × 1024 = 584 sq.ft.

(Contour Interval) × 1 = 584 cu.ft.

3.59 sq.in.

× 1024 = 3680 sq.ft.

× 1 = 3680 cu.ft.

+(40×60×6.5) = 15600

19280 cu.ft.

CORRECTION FOR AREA iCki

= $\frac{0.25 \times 1024 \times 0.8}{3}$ = −68

19212 cu.ft.

(b)   ÷ 27 = 22 cu.yd.   ÷ 27 = 712 cu.yd.

FIGURE 11.8. Computation for cut and fill.

formed outside the rectangle *ABCD* is all fill and within the rectangle we have excavation, a cut. The scale of the plot drawing is 1 in. = 32 ft 0 in. Compute the volumes of cut and fill shown by the contours and excavation area.

Solution: Since the scale of the drawing is 1 in. = 32 ft 0 in., 1 in^2 of area registered by the planimeter is 32 × 32 or 1,024 ft^2, the scale factor. The various areas and the required computations are tabulated in Figure 11.8*b*, the procedure followed being that explained in Section 11.4.

In computing the volumes of cut and fill on contour 89, areas *abca* and *dDefd* are tabulated under "fill" and area *cdAc* (the contours lie in the vertical face of the excavation) under "cut." Cut and fill on contour 88 are recorded similarly. On contour 87, areas *ghig* and *jklj* are fills and *ikDABi* is a cut. The lowest contour intercepting the excavation is 87. Since the bottom of the excavation has an elevation of 80.5, the bottom of the excavation is 87 − 80.5, or 6.5 below contour 87.

Therefore, the volume of excavation not previously accounted for is a prism of earth 40 × 60 × 6.5, or 15,600 ft^3, as shown in Figure 11.8*b*. This method of computation, however, includes an excessive volume; it is the volume *iCki* that is below the 87 contour level. By interpolation, the original elevation of point *C* is 86.2. Consequently, the excessive volume may be considered as a pyramid having *iCki* as a base (0.25 × 1,024) ft^2 and a height of 87 − 86.2, or 0.8 ft. Thus, its volume is 0.25 × 1,024 × 0.8/3, or 68 ft^3. Note that 1,024 is the scale factor for converting square inches on the drawing to square feet. This volume, 68 ft^3, is deducted from the volume of cut previously computed. It is shown in the tabulation in Figure 11.8*b*.

The volumes of cut and fill are 712 yd^3 and 22 yd^3, respectively. As noted in Section 11.4, the volumes computed by this method are approximate. If the excavation is a relatively large part of the total cut, the procedure explained in Section 11.2 should be used for determining the volume of cut within the excavation lines.

12

DRAINAGE AND GRADING

12.1 PROVISION FOR DRAINAGE

Rain water that falls on the surface of a property either evaporates, percolates into the soil, flows off the site, or drains to some point or points on the site. That portion that does not enter the soil is called the *runoff*, and provision must be made for this excess water. The grading must be so designed that surface water will flow away *from the buildings*. This may necessitate drainage channels with catch basins and a system of underground piping.

When grading is required adjacent to existing trees, the natural elevations should be disturbed as little as possible. The elevation should never be lowered more than 6 in. If the ground surface must be raised, stone or brick tree-walls should be provided. They should never exceed 4 ft in height.

12.2 LAWNS AND SEEDED AREAS

The preferred grade for lawns or seeded slopes adjacent to buildings is 2%; the minimum is 1%. Earth banks should have a 1:2 maximum slope, and, if power lawn mowers are to be used, the slope should not exceed 1:3. Water flowing over banks tends to erode the surface and wash out planting. To avoid this, drainage gutters may be constructed at the top of the bank to intercept the water.

12.3 WALKS AND PATHS

It is desirable that walks and paths have a crowned cross section so that surface water is diverted to both sides. In many instances this is impossible and small quantities are permitted to flow across the pathway. For these walks, as well as for pavements adjacent to buildings, the cross slope should have a grade of 2% or 3%. To provide for drainage, a 1% longitudinal slope is the preferred minimum. In cold climates, where ice forms readily, 6% is considered to be the maximum longitudinal slope but, in milder climates, the maximum may be as much as 8%. In general, long walks should have minimum slopes, but for short walks or short sections of walks the slope may, if necessary, approach the maximum. Steep slopes should always be avoided near the entrances to buildings.

Main entrance walks to residences should have a minimum width of 3 ft; for public or semi-public buildings, a 5 ft width is considered to be the minimum. Intersecting paths should have slanted or rounded corners with a minimum radius of 6 ft.

When possible, steps should be avoided in walks. When they are unavoidable, not less than three risers should be constructed since a smaller number may serve as a stumbling block. If there are five or more risers, handrails should always be provided.

12.4 ROADS AND DRIVEWAYS

For drainage, roads and driveways should have a minimum longitudinal slope of 0.5% but a 1% minimum is preferable. A 6% slope is considered to be the maximum, but for short distances it may be as great as 8% or even 10% in mild climates where icy roads present no problem. At road intersections the grade should not exceed 3%.

Concrete and bituminuous-surfaced roads are generally parabolic in cross section and should have a crown of $\frac{1}{4}$ in. for each foot of half-width. For earth roads the height of the crown should be increased to $\frac{1}{2}$ in. per foot of half-width. A dished cross section formed by an inverted parabola is sometimes used for driveways. They are recommended only when concrete is used, and the inverted crown should be $\frac{1}{2}$ in. per foot of half-width.

The width of a road is determined by the number of lanes of traffic and the parking requirements. Each traffic lane should have a 10 ft minimum width. Parallel parking requires a width of 8 ft, diagonal parking a minimum of 15 ft (preferably 18 ft), and perpendicular parking a minimum of 19 ft.

When curves occur in driveways on a property, the radius of the curve to the inside edge should not be less than 20 ft. Streets on

which the traffic speed is limited to 20 mph should have curves of at least 100 ft radii to the inner edge. Intersecting streets should have rounded corners with a minimum radius of 15 ft.

To avoid maintenance expense for the property owners, streets in housing developments are commonly dedicated to the city or township upon completion of the project. To ensure their acceptance, care should be taken that the specifications and requirements of the local highway department are rigidly adhered to in the construction of such streets.

12.5 INTENSITY OF RAINFALL

In the design of a drainage system, one of the primary factors is the expected number of inches of rainfall per hour in a given locality. For drainage systems of comparatively small areas, the maximum rainfall in any 2-year period is generally used but, for a more conservative design, the 5-year period may be employed. Data relating to the volume of rainfall may often be obtained from the records kept by municipalities. When such information is not available the two charts shown in Figure 12.1, prepared by the U.S. Department of Agriculture, may be used. They give the 1-hour rainfall in inches to be expected in the 2- and 5-year periods. One inch of rainfall per hour is equal to approximately 1 ft^3 (actually 1.0083) of water falling on 1 acre of ground per second.

12.6 RUNOFF

All the rain falling on the earth's surface does not reach the drainage lines. Some is lost by evaporation, and some, depending on the porosity of the ground, seeps into the soil. The water that reaches the drainage system is called the *runoff*. Table 12.1 gives average ratios of runoff to the total amount of rainfall that falls on various surfaces.

The volume of runoff may be estimated by

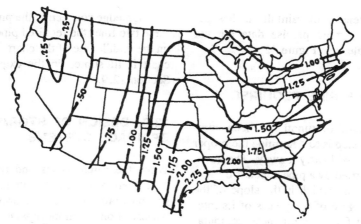

ONE HOUR RAINFALL, IN INCHES TO BE
EXPECTED ONCE IN 2 YEARS

ONE HOUR RAINFALL, IN INCHES TO BE
EXPECTED ONCE IN 5 YEARS

FIGURE 12.1. Zones of rainfall intensity for the United States. From *Miscellaneous Publication No. 204*, U.S. Department of Agriculture.

TABLE 12.1 RUNOFF COEFFICIENTS

Roofs	0.95
Concrete or asphalt roads and pavements	0.95
Bituminous macadam roads	0.80
Gravel areas and walks	
loose	0.30
compact	0.70
Vacant lots, unpaved streets	
light plant growth	0.60
no plant growth	0.75
Lawns, parks, golf courses	0.35
Wooded areas	0.20

the following formula known as the *rational formula:*

$$Q = ACI$$

in which

Q = the runoff from an area, ft^3 per second

A = the area to be drained, acres

C = the coefficient of runoff. See Table 12.1.

I = the intensity of rainfall, inches per hour. If more precise data are not available, use Figure 12.1.

12.7 SIZE OF PIPE REQUIRED

When the volume of runoff has been computed, the next step is to determine the proper size of pipe that will carry it away. The quantity of water carried by a pipe depends on several factors among which is the slope of the pipe and the degree of roughness of its interior surface. In formulas for these computations the term n, the friction factor, is generally taken to be 0.015 for vitrified sewer or drainage pipe. A convenient method of determining the required size of pipe is to use Figure 12.2, an alignment chart. In this chart the left-hand line gives the volume of runoff and the right-hand line gives various pipe slopes for a friction value of $n = 0.015$. By placing

a straightedge connecting the proper values on these two lines the required pipe size is found on the middle line of the chart. The use of the chart is illustrated in the example given in Section 12.9.

12.8 DESIGN OF STORM DRAINAGE SYSTEMS

When the catch basins and roof-drain lines have been located on the site plan, the underground drainage lines are drawn. This may consist of one or more separate systems; they may discharge into a gutter, stream, or public sewer. The drainage system adopted is that which contains the least total length of pipe. Lines should be run with as few bends as possible, using 45° and Y bends at intersections.

The *invert* is the lowest part of the internal surface of a pipe. When the locations of the drainage lines have been established, the elevations of the inverts are computed and marked on the drawing. The pipes should be laid below the frost line and should have a minimum of 3 ft of cover to protect them from the weight of passing vehicles. Lines run parallel to the surface of the ground require a minimum of excavation.

Beginning at the upper ends of the lines, the volume of water at the catch basin is determined and the pipe size is established by the use of the chart shown in Figure 12.2. Water flowing from adjacent properties must be provided for. The drainage line is followed to its point of discharge, accumulating drainage water from other catch basins or branch lines, and the proper pipe sizes established for each additional increase in water volume.

In estimating the volume of water flowing from adjacent properties, consideration must be given to the possibility that vacant land may be built up at a later date.

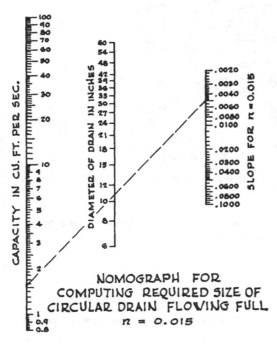

NOMOGRAPH FOR COMPUTING REQUIRED SIZE OF CIRCULAR DRAIN FLOWING FULL
$n = 0.015$

FIGURE 12.2. Nomograph for computing required size of a circular drain flowing full. Adapted from *Engineering Manual*, U.S. Army Corps of Engineers, Dec. 1945.

Example. A 200 ft × 300 ft plot, located in the southern part of Ohio, contains 4,360 ft^2 of buildings, 6,500 ft^2 of bituminous macadam roads, 2,200 ft^2 of concrete walks, and the remainder, 46,940 ft^2, is a seeded and planted area. An adjacent plot, having an area of $\frac{1}{3}$ acre, is vacant land the runoff from which will flow onto the 200 ft × 300 ft plot. Determine the size of the drainage pipe to accommodate the runoff from this plot if the pipe has a slope of 0.005.

Solution: The first step is to compute the volume of runoff; therefore, we will use the formula $Q = ACI$ given in Section 12.7 and the C coefficients for runoff given in Table 12.1. Then,

	Areas (ft²)		Coefficient		Adjusted Area (ft²)
Roofs	4,360	×	0.95	=	4,150
Macadam roads	6,500	×	0.80	=	5,200
Concrete walks	2,200	×	0.95	=	2,100
Lawns	46,940	×	0.35	=	16,430
	60,000 ft²				27,880 ft²

Since there are 43,560 ft² in an acre, 27,800 ÷ 43,560 = 0.64 acre, the value of $A \times C$ in the formula exclusive of the adjoining $\frac{1}{3}$-acre vacant plot. This land may be built up at a later date; consequently the C coefficient will be taken as 0.75. Then, 0.33 × 0.75 = 0.25 and 0.64 + 0.25 = 0.89, the value of $A \times C$.

The intensity of rainfall, I in the formula, is found in Figure 12.1 to be 1.75 in., the 1-hour rainfall expected once in 5 years. Then $Q = ACI$, or $Q = 0.89 \times 1.75 = 1.56$ ft³ per sec, the volume of runoff from the area to be drained.

Now refer to Figure 12.2. Place a straightedge so that it connects 1.56 on the left-hand line with 0.005 on the right-hand line as shown on the chart. The straightedge now crosses the middle line between 10 in. and 12 in.. Therefore, we will accept a 12 in. diameter drainage pipe.

PROBLEM 12.8.A.

A 300 ft by 320 ft plot, located in the western part of Montana, contains 6000 ft² of buildings, 9000 ft² of bituminous macadam roads, 2500 ft² of concrete pavement, and the remainder is seeded and planted. There is also a drainage onto the site from an adjacent plot of $\frac{1}{2}$ acre of vacant land. Determine the size of the drainage pipe to accommodate the runoff from this plot if the pipe has a slope of 0.008.

12.9 SEWER AND DRAINAGE PIPE

Because of its comparatively low cost, small sewer and drainage lines are usually constructed of vitrified-clay pipe. Inside buildings, and within 10 ft of them, only cast-iron pipe should be used. When drainage pipes exceed 15 in. in diameter, concrete and asbestos cement pipe are often used. Owing to the fact that roots frequently clog clay pipes, 6 in. is considered to be the minimum size although some authorities recommend not less than 8 in.. For house drains, cast-iron pipe 4 in. in diameter is commonly used.

12.10 PARKING AREAS

With the increasing number of problems that result from traffic congestion, it has become desirable to provide parking areas on the building site of all but the smallest buildings. These areas are for the accommodation of the occupants as well as for visitors to the buildings. Many municipalities require on-site parking areas for certain types of buildings. In arranging the parking spaces, they should be so arranged that parked cars will not obscure the vision at street intersections or curves.

Care should be taken to see that parking areas are properly drained, the minimum and maximum slope being 0.5% and 4%, respectively. Surface water should not be permitted to flow directly onto public highways but should be intercepted by catch basins or inlets.

12.11 SITE CONSTRUCTION AND DRAINAGE

Site construction of any kind, including buildings, generally affects the character of drainage on the site. This is obvious in terms of the definition of the site surface geometry, but can also relate to various other considerations. For example, the total runoff, as discussed in Sections 12.6 and 12.8, is related to the character of the surface being drained.

General handling of water on a site typically requires some special concerns for construction. One example is the possible effect on basements. It is common procedure to slope the site surface immediately adjacent to a building away from the building to avoid ponding of water at the building edge. This is of greater concern when there is a basement and the possibility of water instrusion through the basement walls.

Site construction intended primarily for other purposes may also serve, deliberately or unavoidably, as drainage control devices. Curbs, retaining walls, and long, constructed planter edging typically serve major roles in directing drainage. Paved walks and driveways, especially ones with high curbs, may serve as drainage channels during periods of intense precipitation.

There are, however, many special forms of construction used primarily for some form of drainage control. Sewer piping, open drainage channels, catch basins, area drains, and curb drains are the most common elements. Some additional elements are described in the following sections.

12.12 SUBSURFACE AND PERIMETER DRAINS

In addition to surface drainage, it is sometimes necessary to have drainage of water from within the ground mass. Reasons for this include the following.

Prevention of Water Intrusion

This is the most common for deep basement levels where excessive precipitation or continuous irrigation can produce considerable water saturation of the near-surface ground mass. A common device used for building basements is the *perimeter drain* or *footing drain*, as shown in Figure 12.3a, consisting of a tile-formed sewer line with open joints surrounding the building just below the basement floor level.

Prevention of Water-Soaked, Retained Soil Masses

These may occur outside basement walls or at the backs of retaining walls and add considerably to lateral force effects against the walls. The perimeter drain will also relieve this for basements. For retaining walls it is also possible to use a linear, open-joint tile drain behind the wall. However, if the drainage is not intrusive, the simpler method for retaining walls is to provide drain holes (called *weeps*) through the wall just above the grade on the low side of the wall.

Prevention of Surface-Level Water Buildup in the Soil

This can occur in open ground and affect the use of areas such as baseball dia-

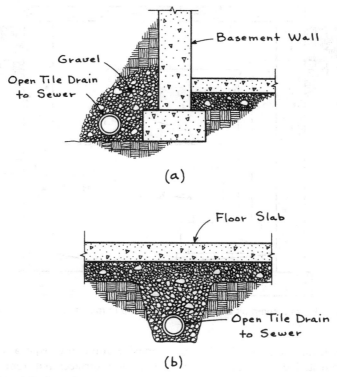

FIGURE 12.3. Use of tile drains with open joints: (a) perimeter drain for basement; (b) subsurface drain in trench.

monds, golf greens, football fields, or running tracks—or, simply, any area where ponding of water is a problem. It can also occur under pavements, including those for building floors.

For the prevention of surface water buildup, the simplest procedure is to use fast-draining soils (mostly sand and gravel with little fine material) for the near-surface soil mass beneath the topsoil or pavement. This may suffice to pull the water buildup sufficiently below the surface so as to prevent the excessive wetting of the surface soils or the underside of the pavement.

Where drainage within the soil mass is not possible or sufficient, it may be necessary to use an undersurface (or underfloor) drain system, with open tile lines similar to those used for perimeter drains, as shown in Figure 12.3*b*.

Perimeter and subsurface drains must, of course, drain *to* something for disposal of the drained water. Their installation necessarily assumes that there is a sewer line for storm water drainage below the level of the open-tile drain system. If not, it may be necessary to utilize elements such as those described in the following two sections.

12.13 SUMPS

It sometimes happens that the development for a site unavoidably places the sewer lines for the site below the level at which the drained materials can be carried away for disposal. A device sometimes used in this situation is a *sump pit* or *sump well*. In its simplest form, this consists of the elements shown in Figure 12.4.

Sumps may be placed completely below

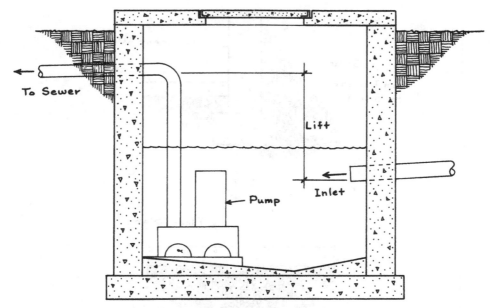

FIGURE 12.4. Use of a sump pit and submersible pump to raise the level of gravity drain line.

ground or have their tops at ground level—or, in some cases, be installed below the floor in a building. In any event, they must be accessible for maintenance and repair, including frequent cleaning of accumulated solids that settle to the bottom of the pit.

Most sumps empty into a sewer main, serving simply to raise the level of the drainage flow to one that can achieve drainage by gravity to the sewer main. They may be required simply because of the great distance to the main and the need for adequate slope of the drainage lines. However, they may also be required because construction places something needing drainage at a very low position, as in the case of very deep, multiple-level basements.

While delivery to a sewer main is most common, flow out of the sump may also be delivered to an open channel or to one of the devices described in the next section.

12.14 TILE FIELDS AND DRY WELLS

Sites in rural locations are often not served by central sewer systems. This requires other considerations for disposal of drained water, both from surfaces and from any piped drains, such as those described in the two preceding sections.

Storm drainage (as opposed to sanitary sewers for toilets) may be simply disposed of at low elevations on a site, as long as no damage to construction or to neighboring properties occurs. If buildings and critical surfaces needing drainage can be placed on higher elevations on the site, this is the simplest method. Delivery may be by piped sewer lines or simply by surface or channelled drainage.

Occasionally, however, it is necessary to dispose of drainage without the use of surface accumulation, that is, within the subsurface soil mass. This generally requires, for one condition, a general soil mass with some degree of permeability that makes the water dispersal feasible. It also requires that the natural groundwater level be some distance below the ground surface, and indeed below the point where actual disposal is intended. You can not add water to ground already saturated.

Two basic devices for disposal of water within the ground mass are *open-tile drain fields* and *dry wells*. The open-tile field is simply the reverse of a perimeter or subsurface drain system—with open-tile joints facilitating flow *out* of the piping lines. The lines of the drain field are simply splayed and spread out in sufficient total ground area to permit the anticipated flow to be absorbed by the ground mass.

The dry well typically consists simply of a hole filled with rocks, although its actual construction is controlled in terms of the total volume and the selection and placement of materials. In due time, the well will fill up with fine materials in the forms of sediments borne by the drained water, so its effectiveness is predictably short-lived. Its life can be extended, however, by making it sufficiently large and by careful control of the gradation of the rocks and gravel used to achieve its highly permeable mass.

A dry well can also be constructed in the form of an open well, with its sides laid up with rocks with no mortar (loose-laid). This is usually much more expensive construction, but permits some extension of its life, as the well can be cleaned periodically of the fine sediments.

13

STAKING OUT SITE WORK

13.1 LAYING OFF ANGLES

The preferred surveying instrument to use in staking out buildings is one in which the telescope may be moved up and down in a vertical plane. This is found in the builders' transit level and the engineers' transit. If the builders' level is used, the plumb bob is employed for transferring levels to points on the ground.

To stake out a building, a surveying instrument is required to lay off given angles. Suppose that we are given the line AB, (see Figure 13.1a), and that we are required to locate point C, 125 ft from A, angle BAC being 42°30′. To begin, the instrument is set up over point A and the 0 mark on the horizontal circle is sighted on point B, as described in Section 5.3. Next, turn the instrument clockwise 42°30′ and at 125 ft from A drive a stake. By directions from the instrument man a tack is driven in the stake marking the location of point C.

Angle BAC is now "measured by repeti-

tion," as explained in Section 5.3. Suppose that we now find the angle to be 42°28′30″. This shows that there was an error of 0°1′30″ in setting the tack; it should have been placed a certain distance farther to the right. Now let us compute this distance. By referring to a table of natural tangents, we find that the tangent of 1′ is 0.0003. Therefore, for each minute of error multiply the length AC by 0.0003 and for each second of error multiply the length AC by 0.000005. Hence, for this problem in which the error is 0°1′30″,

$$1 \times 0.0003 = 0.00030$$

$$30 \times 0.000005 = \frac{0.00015}{0.00045}$$

Since AC is 125′, the total error is 0.00045 × 125, or 0.056′.

Figure 13.1b shows the stake driven at point C, and T_1 indicates the position of the first tack. Since we have found that the exact position of point C is 0.056 ft farther to the

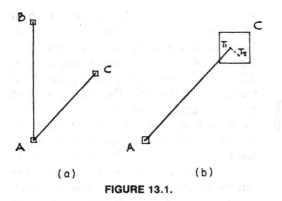

FIGURE 13.1.

The site plan shows the positions of the boundaries and their distances from the proposed buildings. The builder locates the boundaries of the plot, and from the data shown on the plot lays out the building by placing stakes at the corners. This will result in one or more rectangles, as shown by the area *ABCDEFGH* in Figure 13.2. To check the accuracy of the work, the two diagonals of the rectangles can be measured. If there is no error, the diagonals will be equal in length.

right, we measure this distance (on a line at right angles with *AC*) and drive a second tack, T_2, thus establishing the corrected position of point *C*. The first tack should now be removed. For problems like this, in which the angle for the first tack was found to be in error, care must be taken to measure off the distance needed to correct the error in the proper direction.

13.2 STAKING OUT BUILDINGS

The boundaries of building plots should be established by monuments (markers) at the corners, set in place by a registered surveyor.

13.3 BATTER BOARDS

The stakes locating the corners of the building, referred to in Section 13.2, will be displaced during the excavation operations. In order that their locations may be retained, *batter boards* are used. They are placed at sufficient distances (generally 5 ft or 10 ft) outside the perimeter of the structure so that they will not be disturbed by the building operations.

Batter boards consist of vertical stakes (2 in. × 4 in.'s or 4 in. × 4 in.'s) driven firmly into the ground with 1 in. or 2 in. boards mailed across them, as shown in Figure 13.3. Notches or saw cuts are cut in the boards so

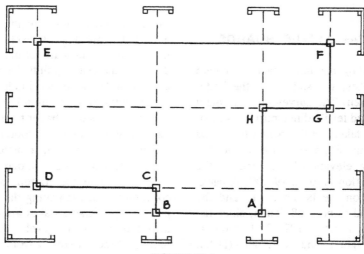

FIGURE 13.2.

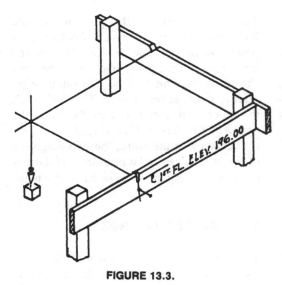

FIGURE 13.3.

that cords or wires may be strung from one batter board to another. The intersections of the cords are directly above the corners of the building. The batter boards should be at the same height, and this height should coincide with a floor level or bear some definite relation to it. The elevation should be plainly marked on the batter boards as indicated in the figure. The elevations of the batter boards should be checked frequently during the excavation and foundation work to see that they have not shifted or been displaced.

13.4 SETTING BATTER BOARDS

After the building has been staked out with the corner stakes, the stakes for the batter boards are driven. The surveying instrument is now set up and levelled in a position so that sights may be taken on the bench mark and the batter boards. Suppose, for example, the bench mark is at elevation 197.53 and that the first-floor elevation is to be 196.00. A backsight is taken on the bench mark, and the reading is found to be 4.68. Then, since the bench mark is 197.53, 197.53 + 4.68 = 202.21, the height of the instrument (H.I.).

The difference between H.I. and the first-floor elevation is 202.21 − 196.00, or 6.21 ft. These heights and elevations are shown in Figure 13.4. The target on the rod is set at 6.21′, and the rod is placed alongside one of the batter-board stakes. The rod is moved up and down until the horizontal cross hair coincides with the target. The height of the bottom of the rod is then marked on the stake, and batter boards are nailed to the stakes so that their tops are level with this mark. The carpenters' level is used to level the batter boards. This procedure is repeated for each set of batter boards.

The instrument is now set up over one of the stakes marking a corner of the building, and the lines of the building are projected to locate the saw cuts and notches on the batter boards. For example, the instrument is set up over point *A*, (see Figure 13.2) and sighted on point *B*. This establishes line of sight *AB*, and a saw cut is placed on the batter board. A less accurate but more rapid method would be to stand to the right of point *A* and to sight-in the point on the batter board by eye. By this method points may be located with an error generally not greater than $\frac{1}{8}$ in.

13.5 LAYING OUT COLUMN FOOTINGS

When the excavations for the building have extended to the lowest level, the centers of the column footings are located and stakes are driven at the center points. The marks on the batter boards may be used in locating the column centers. If the column footing is large, it may require its own batter boards to establish its perimeter. For small footings, stakes are driven in the shape of a rectangle a whole number of feet (2 ft or 3 ft depending on conditions) outside the footing perimeter. For convenience in excavating the footing to the proper depth, these stakes are driven so that their tops are at the level of the top of the footing. Lines stretched from stake to stake

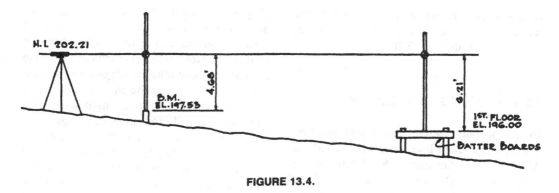

FIGURE 13.4.

may be used to locate the centers of the column footings, and lines are also used to establish the perimeter of the footing.

13.6 STAKING OUT ROADS AND DRIVEWAYS

In laying out roads and driveways, the usual procedure is to drive stakes along the center lines. These stakes are located at each point of change of slope and at regular intervals, 25 ft or 50 ft, along the length of the drive. The elevations of the various points are taken from the site plan. The stakes are driven so that their tops are at the elevation of the finished driveway, or at a number of full feet above the required elevation, and are so marked. They are first driven with their tops several inches above the desired elevation. The rod, with the target placed at the required setting, is held on top of the stake, and the stake is then driven into the ground until the center of the target coincides with the level line of sight of the instrument. On straight runs of the road, stakes may be located on "offset lines" a number of full feet away from and parallel to the center line of the road. These stakes are so located that they will not be disturbed during the construction operations. The methods used for locating points on circular and vertical curves are given in Chapters 7 and 10, respectively.

13.7 GRADES OF CONSTANT SLOPE

When a number of intermediate stakes are to be set on a portion of a road that has a constant slope, the following method may be used: Stakes are first located and driven to the required elevations at the two ends of the slope. The instrument is now set up over one of these stakes, and the height from the stake to the center line of the telescope is measured. The target on the rod is then set at this measured height, and the rod is held over the stake at the other end of the constant slope. Now the telescope is tilted so that the horizontal cross hair coincides with the target. The line of sight of the instrument is now parallel with the slope of the road. Without altering the position of the target, the rod is then held over the successive intermediate stakes, which are now driven to a depth that permits the center of the target to coincide with the line of sight of the instrument.

13.8 SETTING STAKES FOR GRADING

When an area of a plot is to be graded to establish new elevations for the surface of the ground, an adequate number of stakes are driven to aid in establishing the new elevations. It is customary to drive these stakes so that their tops are a full number of feet above or below the finished elevations. After they are driven they should be plainly marked, such as ''4 ft fill'' or ''2 ft cut.''

13.9 STAKING OUT SEWERS

In staking out a sewer line, stakes are first driven on its center line at regular intervals of 25 ft to 50 ft. On each side of this line of stakes, far enough apart to prevent them from being disturbed by the trench excavation, heavy stakes are driven to which planks are nailed across the sewer line. Small boards, placed vertically, are nailed to the planks with one edge in alignment with the center line of the sewer. A nail is now partially driven into the vertical board at some full number of feet above the *invert* of the sewer pipe. The invert is the lowest part of the internal surface of the pipe. Cords or wires stretched from nail to nail mark a line at some established height above the invert, and a vertical board held beside the string is used to lay the sewer pipe at the desired elevation. The drawings showing the sewer lines should include a profile of the sewer, indicating its elevations at various points and their relation to the surface of the ground.

13.10 GENERAL EXCAVATION

On sites where construction is planned over a considerable area, a common procedure is to first perform a *general excavation* and grading. This may consist of a single, flat level or of several separate, flat portions in a terraced layout. If some filling is involved, it may be more accurately described as a *general preliminary grading*.

The general purpose of this work is to sim-plify the staking out and excavation for foundation construction of individual elements of construction—usually individual buildings on a large, multibuilding development. It is also a matter of more efficient staging of the construction work—using the equipment required for extensive grading in a single action, instead of repeatedly for single buildings.

For this staged operation, there are two different events of staking out the site. The first serves to instruct the workers in the grading and excavating required for the general site profiling. The second deals with individual preparations for separate elements on the site. Obviously, some greater concern for accuracy is present for the second stage in this case.

13.11 FINISHED GRADING

Upon completion of construction work, the final site development event is the finish grading. This consists generally of the final surfacing of open ground areas, typically in preparation for the installation of plantings.

For sites with large open ground areas, this may require extensive staking out—especially if complex, rolling, or terraced ground forms are planned. In other cases, with minor open areas, the installed construction (buildings, curbs, etc.) may effectively define the levels for the open ground areas, requiring little or no staking.

For large sites the finish grading may be accomplished in two stages. As for a general preliminary grading, the approximate finish grade may be established by a general grading contractor. The actual finish surface, however, is likely to be developed with topsoil by a landscaping contractor, in preparation for the installation of general ground cover plantings, such as grass lawns.

14

SITE STRUCTURES

Development of a site may consist primarily of the reshaping of the ground surface and possibly the replacement of some surface materials and introduction of new plantings. However, various forms of structures may also be required. A building will constitute a major construction, but various other structures may also be used for the site development. This chapter describes forms of site construction other than buildings. Construction of elements involves various materials, typically stone, masonry, concrete, wood, and asphalt. This book deals with the various site development concerns of these materials, rather than their actual structural design.

14.1 PAVEMENTS

Paved surfaces may be achieved with various materials, the common ones being the following (see Figure 14.1).

Concrete

This is usually the strongest form of paving and can be used for heavy traffic-bearing roads or simple walks. Figure 14.1a shows a simple paving slab of reinforced concrete, typically placed over a thin layer of gravel or crushed rock to provide a good base and a draining layer beneath the pavement. Minimal reinforcement may be provided with a single layer of steel wire fabric; thicker pavements may be reinforced with two-way grids of steel rods.

Asphalt

Various forms of concrete produced with an oil-based binder (tar, etc.) can be used. Materials, thickness, and base preparation depend on the desired degree of permanence and cost limitations.

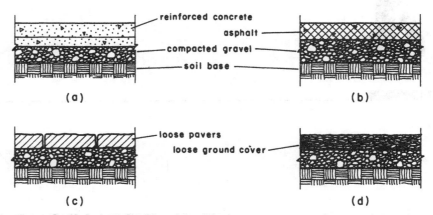

FIGURE 14.1. Forms of paving: (a) concrete slab; (b) asphalt over gravel; (c) loose pavers; (d) compacted loose material.

Loose Unit Pavers

Bricks, cobblestones, cast concrete units, cut wood sections, or other elements may be laid over a sand and gravel base. This is an ancient form of paving, and it can be very durable if installed properly and made with durable elements.

Loose Materials

Gravel or pulverized bark may be used for walks or areas with light traffic. These generally require some ongoing maintenance to preserve the surface, but can be very practical and blend well with natural features of a site, such as existing surfaces, plantings, and so on.

Areas to be paved must be graded (recontoured) to some level below the desired finished surface to allow for pavement construction. If existing site materials are undesirable for the pavement base, it may be necessary to cut down to a lower surface elevation and to import materials to build up a better base for the pavement.

Pavements, especially of the solid form of concrete or asphalt, result in considerable surface runoff during rainfall, which must be carefully considered in the general investigation of site surface drainage. (See Chapter 12.)

14.2 RETAINING STRUCTURES

Site development frequently involves the use of various retaining structures. These help to achieve abrupt changes in the ground surface elevation. The form of structure often relates primarily to the height of the elevation change on the two sides of the retaining structure.

The smallest retaining structures are curbs, which may take various forms, as shown in Figure 14.2. Curbs are usually edging devices that define the boundary between different units of the site surface. They are frequently placed at the edges of pavements and thus often relate to the paving materials and forms. The form of drainage of the pavement may also affect the details of the curb.

Curbs are usually limited to achieving elevation changes of less than 18 in. or so. As the height of the retaining structure increases, a different general form of construction is required.

14.3 LOOSE-LAID RETAINING WALLS

For abrupt elevation changes of more than 18 in., some form of wall construction is required. This may be achieved with loosed-laid stones (without mortar) or other elements, as shown in Figure 14.3. Such construction must

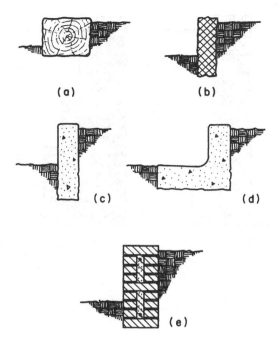

FIGURE 14.2. Curbs: (a) timber; (b) stone; (c) concrete—precast or sitecast; (d) cast concrete with gutter; (e) masonry.

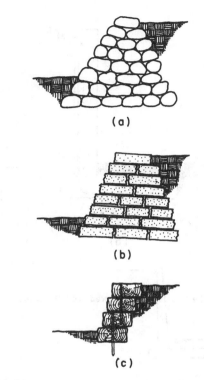

FIGURE 14.3. Loose-laid retaining walls: (a) field stone; (b) broken concrete; (c) timber.

be banked, or leaned, into the cut to resist the horizontal force of the soil on the high side of the wall.

Walls of this type can be very effective and simple to construct. Executed with field stone or timber they may be quite attractive, especially when used with other largely natural materials, such as earthen surfaces or plantings.

A typical advantage of the loose-laid wall is its natural porosity, which allows groundwater seepage. This is especially critical in the event of regular heavy irrigation of plantings on the high side.

14.4 CANTILEVER RETAINING WALLS

The strongest retaining structures for achieving abrupt elevation changes are *cantilever retaining walls*, which consist of structural walls of masonry or concrete anchored to a

large footing. Some common forms are those shown in Fig. 14.4.

A form used for walls up to 6 ft in height is shown in Figure 14.4a. (Height refers to the ground surface elevation difference on the two sides of the wall.) Critical structural concerns are for the rotational, overturning effect and the horizontal sliding effect, both caused by the horizontal soil pressure on the high side of the wall. The dropped portion of the footing (called a shear key) is commonly used to enhance the resistance to horizontal sliding.

Short walls may be achieved with masonry or concrete. If achieved with concrete, both the wall and the footing are usually reinforced as shown in Figure 14.4a. Masonry walls may be constructed similarly, as shown in Figure 14.4b, or may be developed as gravity walls, relying strictly on their dead weight to resist the overturning effects, as shown in Figure 14.4c.

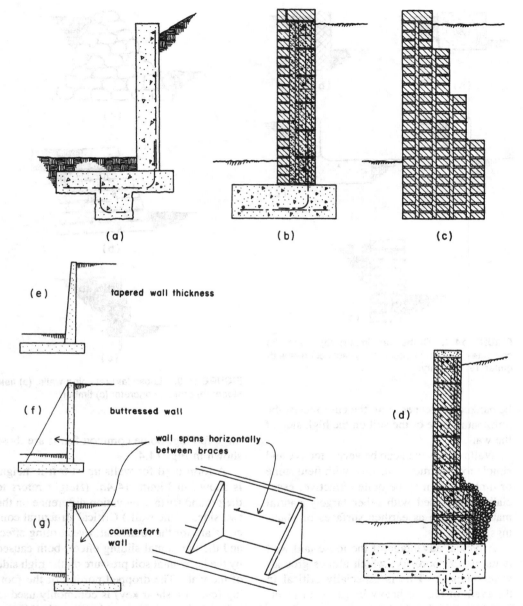

FIGURE 14.4. Concrete and masonry retaining walls: (a) short concrete wall; (b) short masonry wall with reinforced CMUs; (c) brick gravity wall; (d) tall concrete wall; (e) step-tapered masonry wall; (f) and (g) braced walls.

As walls get taller, it is common to use a tapered wall form. Concrete walls are evenly tapered, as shown in Figure 14.4d, while masonry walls are typically step-tapered, using regular units of the masonry, as shown in Figure 14.4e.

Tall retaining walls may also be braced by buttresses or fin walls perpendicular to the retaining wall, as shown in Figure 14.4f and g. If built on the back side, these do not affect the viewed side of the wall (although building them on the exposed, low side is usually easier and more economical, as it involves less excavation and backfill on the high side).

Retaining walls may also be developed as part of building construction, becoming building walls with some bracing of the wall often provided by other elements of the building construction. The typical basement wall is generally a retaining structure, although not usually of the cantilever type, as it works by spanning between the basement and first floor constructions that brace it laterally.

14.5 CHANNELS

Ground surface drainage tends to become focused into channels. Following natural ground contours, natural channels eventually collect to form creeks and rivers. For developed sites it is usually necessary to form new channels as part of the design for surface drainage.

Constructed channels may be closed, in the form of piping or tunnels for sewers. In some cases, however, open channels may be formed to feed into existing sewer systems or into natural rivers or lakes. Open channels may be formed basically as ground forms, but usually have some form of lining of the bottom and/ or sides to prevent washouts and erosion over time.

Fig. 14.5 shows the forms for various types of open channels. For landscaped sites, the channels may be developed as natural-appearing streams, and, in keeping with other site construction materials, some appropriate, attractive form of construction may be used for the channels.

The strongest channels are usually those made with reinforced concrete construction, such as that shown in Figure 14.5c. These may be used to ensure more positive control against erosion, or may be required where major flow is anticipated due to local rain storm conditions.

Constructed channels are most often part of some general region flood control or storm drainage system. They may already exist on a site, even if other development has not been achieved. In any event, they usually have

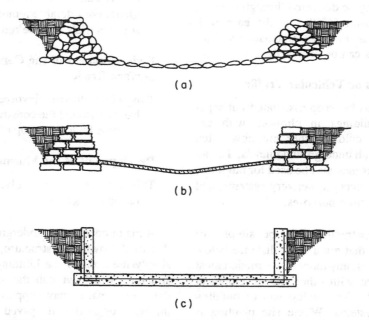

(a)

(b)

(c)

FIGURE 14.5. Construction of open channels: (a) field stone walls and bottom; (b) broken concrete walls and asphalt bottom; (c) reinforced, sitecast concrete walls and bottom.

connections off the individual site, and their design, alteration, and utilization are subject to various controls.

14.6 TUNNELS

Site development for large projects usually involves some use of underground tunnels. Some examples of tunnel use are given below.

Sewers

Small sewers usually consist of buried piping, which can be made of steel, cast iron, fiber-reinforced plastic, or fired clay. As the required flow capacity increases, larger pipe may be used, but some form of tunnel construction can also be used.

Utilities

For various reasons, services such as gas, water, electrical power, and telephone lines may be delivered through tunnels. One reason for this is the ease with which continuous maintenance and alterations can be achieved.

Pedestrian or Vehicular Traffic

Connections between associated but separate buildings in climates with extremely cold weather are now often made with underground tunnels. Tunnel networks may also be used for rail lines, waste collection, delivery systems, and various other purposes.

Tunnel construction may be simple for small tunnels that are a short distance below grade. However, any tunnel construction must be coordinated with other site development work and with the development of building foundation systems. Where site planting is developed, the tops of tunnels must be suffi-

ciently buried to allow for the plant growth over them.

14.7 BELOW-GRADE BUILDINGS

Site surfaces are sometimes developed over building spaces that are completely underground. This may occasionally occur with an entire underground building, but more commonly, only a portion of the extended basement of a building is involved. A frequent occurrence of this type involves the development of underground parking for a building, in which the parking structure may exist below a major portion of the site, while the building above the ground is over only a portion of the site.

In this case, development of the site generally involves three basic concerns, as follows.

Excavation and Construction for the Underground Constructed Building

If this is entirely below the original site grade, considerable removal of existing soil materials may be required.

Redevelopment of the General Finished Surface Grade

This will ordinarily involve considerable backfill around the construction, as well as some buildup on top of the structure.

Development of the Finished Surface

This will generally involve plantings, a paved surface, etc.

A major concern for underground spaces is that of the overhead structure, which is *not* simply the floor of the building above. This structure, which we call the roof of the underground space, may support soil of some thickness or be directly paved over as a terrace or plaza surface. With an extensive underground construction and major landscape development, all possible overhead conditions may exist.

Structure

Compared to ordinary roofs, those for underground spaces usually carry many times the total gravity loads. The construction of the structure itself is likely to be among the heaviest types of sitecast concrete. Add to that either a heavy load of soil or the high live load usually required for a terraced roof area or a plaza (150 to 250 psf in most codes).

A further concern is often that of restricting the total depth of the structure to minimize the depth of the underground construction. All of this conspires to generally keep spans short, with an aim for some efficiency in the spanning systems if possible. The usual options for sitecast concrete floor systems are all possible, although a few tend to be more feasible. Among these, the two-way flat slab and two-way waffle systems usually offer the smoothest underside and least overall depth and are favored, especially if the underside of the structure is left exposed.

Water Control

A roof is a roof, and the general, basically flat roof of an underground structure is not significantly different in many regards from any flat roof. A totally watertight membrane is required, together with all flashing and careful sealing of any penetrations. The membrane must be protected—especially here—and some insulation and a vapor barrier may be indicated.

Drainage is somewhat different here than for a conventional roof, but not any less indicated. The form of drainage will relate somewhat to what is above the roof (soil or paving) and probably to the development of a total drainage system for the whole underground construction. As in other situations, the roof must be drained essentially by grav-

ity flow, involving considerations for the total slopes required and the vertical dimensions of the construction necessary to achieve drainage and water removal.

As for walls and floors, concern for the watertight security of the underground space may be heightened if the space is occupied by people. However, there is hardly any middle ground for a watertight roof—it is truly watertight or it is not, so this is less a variable issue than it is for basement walls and floors in ordinary circumstances.

Earth-Covered Roofs

An especially critical water control problem is that of the roof supporting earth, and usually plantings. The earth may get wet from precipitation alone, but the plantings will require continuous moisture, so the water condition will be ever present. But drainage must be effective to prevent a saturated, rot-inducing condition for the plant roots.

Another major consideration in this situation is the provision of sufficient depth of earth fill for sustaining of the planting. This may be minimal for grass, but of major proportions for trees. To save depth, large plants and trees may be placed in special planters integrated into the space of an otherwise not so deep underground construction.

Figure 14.6 shows some of the features of an earthcovered underground structure, sustaining only minimal plantings in this case.

Paved Roofs

When close to the surface, underground spaces may have paved roofs forming terraces or plazas for buildings, or simply parts of a general site development. Extensive underground parking garages are developed in many locations with parks, squares, or other open spaces on their roofs. Paving here is not essentially different from other outdoor paving, with all the usual options. The sub-base

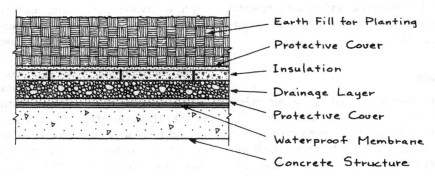

FIGURE 14.6. Typical components of an earth-covered roof.

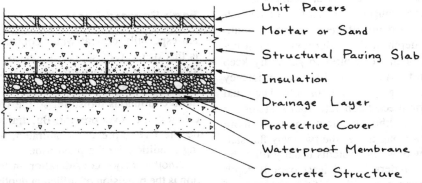

FIGURE 14.7. Typical components for the development of a plaza pavement over an underground structure.

and general support may be somewhat different and must be generally integrated with the overall construction of the roof of the space below. Figure 14.7 shows the typical components of a plaza pavement over an underground structure.

The same underground space may have both earth for plantings and paving on different areas of the roof. If the roof is constructed essentially flat, this calls for some coordination of the overall dimensions for development of the two surface types.

15

MANAGEMENT OF SITE
MATERIALS

An existing site represents an inventory of materials. In the process of developing the site, decisions must be made that involve the management of this inventory. These decisions may involve the removal, rearrangement, modification, or replacement of materials. The nature of the existing site and the type of redevelopment will determine the extent to which site modification is required.

15.1 SITE MATERIALS

Site materials can be broadly divided between those below grade and those above. Below grade are mostly various soil, rock, and water deposits. For previously developed sites there may also be various below-grade constructions. For heavily forested sites, there may be considerable root growth below grade.

Above the ground surface there may be trees or other plant growth. The preservation or removal of these must be determined in conjunction with the site development plans, this is discussed in Chapter 16. If preservation is desired, it may be necessary to develop careful plans to protect existing growth during site work and construction. Raising or lowering the grade, major changes in surface drainage, and other modifications may seriously affect existing growth.

General concerns for soil materials are discussed in Appendix A, which also describes various soil properties and the means of identifying soil types. Soil is basically natural, but modifications are possible—and indeed often necessary— for site development.

In general, existing site materials must be viewed as a given inventory of materials that must be managed in the site development process. The existing materials may be removed, replaced, modified, relocated on the site, or simply left as is.

Problems that may affect the management of site materials include the following.

Establishment of Finished Grades

If the level of the site surface (grade) is to be substantially raised or lowered, there will be major needs for the removal or importation of materials.

Building Excavation

For large buildings with extensive below-grade construction, the extensive removal of site materials may be necessary to excavate for construction. These materials may be used elsewhere on the site, or will require considerable planning for transportation and disposal off site.

Extensive Landscaping

Major new plantings may require the importing of considerable material for surface soils to sustain plant growth.

Site Construction

Extensive development of site structures may require the removal of soils displaced by the construction.

Special problems involving these and other concerns are discussed in the remaining sections of this chapter. The anticipation of these problems will often affect the type and extent of data required from site surveys and subsurface investigations.

15.2 REMOVAL OF SOIL

When possible, it is best to balance the cuts and fills required to achieve finished grades so that no significant removal or importing of soils is required. This is not always possible, however, and some major removal may sometimes be required.

A common situation requiring extensive removal of soil is that of major below-grade construction. Most sites are not raised or lowered significantly simply because they must

retain some connection to the site boundary conditions of neighboring properties or streets. Site construction therefore displaces soils that must be removed.

The other major reason for removal of existing site materials is that they are undesirable for some reason. For plantings, better support of pavements, drainage, or for various reasons, the existing soils may not be usable and or feasibly modifiable.

Removed soils must be taken somewhere, which may present a major problem for the site development, all the more so if they are basically undesirable.

15.3 IMPORTED MATERIALS

In the best of situations, the materials desired for importing to one site may be those required to be removed from another site. Where extensive, ongoing construction occurs, this exchange is frequently made.

Topsoil for plantings required for one site may be removed from another site that is to be mostly covered with buildings, site constructions, and pavements. Or, the soil removed to achieve a major excavation on one site may be used to raise a major depressed portion of another site.

Where this is not the case, the practicality of a particular proposed site development may hinge on the feasibility of obtaining the necessary imported soil materials.

15.4 MODIFICATION OF SITE MATERIALS

Existing site materials often represent usable raw materials that require some modification for the purposes of the site development. Surface soils to be used for surfacing may need to be cleaned of debris, large roots, rocks, and so on. Soils to be used for structural purposes, such as base supports for pavements or bearing supports for foundations, may need some other forms of modification. These latter modifications may seek to improve soil strength, improve resistance to settlement-producing deformations, change water-related properties, or simply improve general soil stability.

Various forms of soil modification and the means for achieving them are discussed in Section A.6 of Appendix A. The feasibility of modifications should be carefully studied before decisions are made about major removal or replacement of site materials.

15.5 SLOPE STABILIZATION

Site development and building construction often requires the creation of some sloping of the ground surface. When this is necessary, a critical decision is required, that regarding the maximum feasible angle for the sloped surface (see Figure 15.1a). Stability of the slope may be in question for a number of reasons. The two principal issues relating to the soil materials are: the potential for erosion from excessive rainfall or runoff of melting snow, and the possibility for sagging or slipping of the soil masses in a downhill movement.

The sloped surface may be created in two different ways: by cutting back of the existing ground, or by building up with fill materials, either imported or borrowed from other site locations. The relative stability of the slope with regard to both erosion and slippage will depend to a large degree on the nature of the soil materials at and near the slope surface. For cut slopes the soil is largely determined, and slope limits must be derived on the basis of the soil properties. For a pure sand, the stable angle is quite easily determined, although erosion is probably quite critical. For rock or some very stable cemented soils, a vertical face—or even a cave—is possible. For more ordinary surface soils, the proper angle must be derived from studies of potential loss of the slope face by various mechanisms of failure.

If the slope is created by regrading the site, and some materials are brought in to create

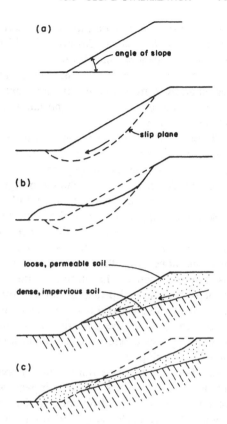

FIGURE 15.1. Considerations for slope failure: (a) measurement of the critical slope angle; (b) typical slope plane failure in loose soils; (c) failure mechanism in stratified soils.

the surface, it may be possible to improve matters by both the selection and the means of installating the new materials. Slope loss by either erosion or slippage is often due to a combination of water soaking of the surface soils and the presence of a relatively loose soil mass. The effects of water on the slope can be controlled by a number of means, including use of less permeable materials.

Erosion is a problem related to many considerations of general site landscaping, drainage, irrigation, and various forms of site constructions (pavings, retaining walls, planters, trenches, ditches, etc.). Erosion control is not usually a foundation problem unless either the loss or buildup of soil masses effects changes in the surcharge effects or lateral forces, or in extreme cases some undermining of bearing footings. The foundation designer should be aware of the general site development as it may affect foundation conditions, but the implementation of measures for erosion control is usually the responsibility of the landscape and site designers.

Slippage of the soil mass on a slope may result from a number of causes, although a very common one is simply the vertical movement due to gravity, as shown in Figure 15.1*b*. In a common mechanism of failure, the soil in the upper portion of the slope simply drops, pushing the soil in the lower portion of the face out into the slope face and down the slope. One theory for the form of this failure visualizes the motion of the soil by rotation along a slip face of curved form, indicated by the dashed-line profile in Figure 15.1*b*. Analyses of the performance of this movement provide one basis for determination of a safe slope angle.

Another type of slope failure that is common is one that occurs due to slippage between two soil strata of different character, as shown in Figure 15.1*c*. A typical situation of this type involves a relative soft, permeable soil mass above a denser, less permeable soil mass. When the upper soil mass becomes saturated, it may move by a combination of ordinary slope failure and slippage along the face between the two soil masses.

For many ordinary situations reasonable slope angles are established by experience and general rules of thumb. For very large projects, long slopes, and any very steep slopes, the entire site and foundation development should be carefully studied by qualified soils engineers armed with extensive data regarding the surface and subsurface soil conditions.

15.6 BRACED CUTS

Shallow excavations can often be made with no provision for the bracing of the sides of the excavation. However, when the cut is quite deep, or the soil has no cohesive character, or

the undercutting of adjacent construction or property is a concern, it may be necessary to provide some form of bracing for the vertical faces of the cut.

One means for bracing is through the use of steel sheet piling (Figure 15.2*b*), which consists of pleated interlocking units. These units are driven individually by pile-driving methods prior to the excavation work and may be withdrawn for reuse or become part of the permanent construction. If the cut is quite deep, it may be necessary to brace the upper portion of the sheet piling as the excavation proceeds.

A simpler form of bracing, used for shallow cuts and cuts where some minor movement of the cut face is not a critical concern, consists of stacked horizontal planks, called *lagging*. These are installed as the excavation proceeds and are usually braced by spaced vertical elements, called *soldier beams*, which in turn must sometimes be braced with struts,

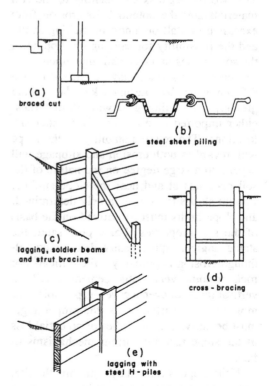

(a)
braced cut

(b)
steel sheet piling

(c)
lagging, soldier beams
and strut bracing

(d)
cross - bracing

(e)
lagging with
steel H - piles

FIGURE 15.2. Elements used for braced excavations.

as shown in Figure 15.2c. If the excavation is narrow (as in a trench), bracing may be achieved with horizontal cross struts that counter-poise the two cut faces (see Figure 15.2d).

For relatively deep cuts or situations where the prevention of movement of the cut face is important, a form of bracing consists of driving piles (usually steel H-shaped piles) for use as soldier beams (see Figure 15.2e). As in the case of sheet piling, these elements may be withdrawn for reuse, or may become part of the permanent wall construction at the cut face.

Bracing is quite commonly required for construction on urban sites where undermining of sidewalks, streets, or adjacent property is critical. The design of such bracing should be closely coordinated with the design of the building substructure. Elements of the bracing system may become parts of the permanent construction, or the interior elements of the subgrade-level building structure may serve as horizontal bracing for the cut faces, replacing temporary bracing elements as the construction proceeds.

Existence of a high water level in the ground or the presence of troublesome soils, such as soft clay, fine sand, or quick silt, can seriously complicate excavation work and make ordinary bracing methods unfeasible.

15.7 WATER CONTROL DURING SITE CONSTRUCTION

Construction work on a site, especially that involving any extensive excavation, must be carried out in reasonably dry conditions. Where the groundwater level is close to the surface, this may require some form of *dewatering*, which generally consists of some means of temporary lowering of the groundwater level.

Dewatering may also be done, however, simply to improve conditions for handling of very wet soils for regrading, building up of slopes, or other purposes. For any reason, dewatering must usually be done so that the groundwater level is lowered several feet below the affected soil.

Dewatering can be achieved by various methods, including ones that employ some means of pumping water from the ground mass and disposing of it somewhere off the affected site. The means for achieving dewatering and disposal of the water will depend on various factors, including the following:

The types of soils involved—slow-draining ones, such as clays, being very difficult to drain.

The extent of area involved and amount of water to be removed (both total volume and rate of flow).

Exactly how and where the pumped water is disposed of.

How long the operation must be in effect to protect construction work.

The simplest form of dewatering consists of creating a pit or trench within the site, below the affected area, and pumping water out as it seeps in. More elaborate systems employ well points extended into the soil mass below the affected area and attached to pumps that lift and remove the water from the soil.

Another form of water control can be achieved with perimeter dams. These may consist of walls of driven sheet steel piling or of slurry trenched walls. Slurry walls are formed by digging a trench and progressively displacing the excavated soil with a bentonite clay slurry. Once installed, when filling the trench the bentonite functions to form a wall within the soil to provide both a structural wall and a water-blocking barrier. If a permanent wall is desired, it can be formed by depositing concrete from the bottom up in the trench, maintaining the structural wall functions during the process.

For very deep excavations, either temporary walls or the actual constructed walls for the building can be used as dams to block the intrusion of water through the sides of the excavation. This reduces the problem to removal of water that enters through the bottom of the excavation as the work proceeds.

16

LANDSCAPING CONSIDERATIONS

This chapter deals briefly with some of the considerations of site development as total landscape, maintaining some significant concerns for other than the simple physical problems of the site. The complete development of a building site often involves a general development of the site landscape. This may vary from a very light trim of the natural site conditions to a completely redeveloped site. There are many different possible arrangements of the division of responsibilities for site work, and who is specifically responsible for the engineering work on the site may shift on different projects. The material in this chapter does not address the issue of assignment of responsibilities, but rather concentrates on engineering problems related to landscape development.

16.1 GENERAL LANDSCAPING WORK

Any landscape architect worth his or her salt would bridle at being seen as dealing only with plantings and other decorative features of sites. The "landscape" is a total experience that relates to all the elements and relationships on a site, including those of the building/site interaction and those extending to the site periphery and its surroundings. Nevertheless, the discussions in this chapter are essentially intended simply to add some concerns for the landscape development that impinge directly on the work of site engineering.

Landscape work typically includes general development of the site surface with plantings, walks, and various site structures. There may also be piping installations for irrigation or fountains and electrical wiring for lighting. Installation of these elements must be coordinated with other work on the site.

One important aspect of coordination of the landscaping work has to do with timing. Some things must be dealt with very early in the design process, and some site work must be delayed until almost the end of the construction process. If existing site features are to be preserved, preliminary landscape design work should begin early enough so that no significant site work occurs before the features to be saved (existing trees, rock out-croppings, streams, etc.) can be adequately protected.

Early surveys of the site (visual, photographic, and instrument-generated) should include the necessary location and identification of landscape features of possible concern. Again, the intention should be to let landscape design concerns be known before work proceeds on site regrading, excavations for construction, placement of temporary construction facilities, and so on. Valuable topsoil materials should be carefully stripped from surfaces and stockpiled for later use.

In the end, of course, final touching up of the landscape is often one of the last stages of the construction process. Good planning should prevent the necessity to perform major regrading or installation of deep-seated structures, piping, or wiring at this stage.

16.2 PROVISIONS FOR PLANTINGS

Plantings occur in a number of forms, the principal ones being as follows:

Existing materials—trees and other forms.
Seeded grass lawns and other forms of general ground cover.
Small plantings—flowers, shrubs, etc., in controlled locations.
Trees and very large shrubs.

Living plants require water, air, protection from physical damage, rooting in nourishing soil, and some controlled relationship to sunlight. Redevelopment of the site surface and materials and general development of site construction must acknowledge these considerations for both existing plantings to be retained and new plantings.

Where considerable landscape development is to occur, the earliest site surveys should include notation and data relating to existing features of likely significance to the landscape design work. Some of these major

concerns are the following:

Existing plant growth of a reasonably healthy character.
Existing features such as large rock outcroppings. Water surface drainage and retention, as related particularly to existing plant materials. (Regrading may seriously imperil their sustained growth.)
Character and extent of existing topsoil with potential for use in new planting development, especially for any extensive lawns.

Lawns may be seeded in topsoil of minimal thickness. However, trees and large shrubs need some considerable volume of soil for root growth. Major excavation may be required for the installation of large trees, making their location a concern for coordination with other site construction, especially that for buried services such as sewer lines.

Large trees may also be expected to produce root growth that will cause problems for pavements and small site walls, and possibly intrusions into piping and tunnels.

In general, it is best to coordinate planning for landscape development with all other site work, from the earliest surveys to the last day of construction work.

16.3 WATER CONTROL FOR LANDSCAPING

Extensive, continued irrigation can cause major changes in the ground moisture condition. This is especially of concern in arid regions, where highly voided soils may have maintained a stability for a long time due to low ground moisture, but can be collapsed easily if saturated.

Regrading may alter the surface drainage pattern on the site, channelling water away from existing growth that has generally thrived under existing conditions. Extensive new pavements may also block both water and air from the roots of existing plants. Major height change of the ground surface may either expose or excessively bury the roots of existing plants. In general, the retention of existing plants means more than simply protecting them during construction work.

New site drainage patterns may produce erosion of existing or new surface soils. Runoff from both precipitation and irrigation must be carefully studied for various concerns, including erosion from planted areas.

Construction below ground must be protected from water intrusion in general if enclosed items are water sensitive. Buried electrical wiring and building basement areas are two examples. Water from extensive, continued irrigation may be critical in these cases, requiring coordination of design. This may involve reconsideration of the use of plantings requiring extensive irrigation or the provision of exceptionally water-resistive construction.

16.4 CONSTRUCTION FOR LANDSCAPING

In a sense, everything on the site is a part of the landscape. However, some items may be specifically created essentially for the landscape development. Planters, fountains, site furniture, site lighting, and edging for planted areas are examples. Provision for this construction must be a part of the site development, affecting the balancing of cut and fill, removal or importing of soil materials, and general development of site drainage.

In a well-coordinated site design effort, some constructions may serve multiple use. Dividing curbs in a parking lot can define edges of planted areas. Retaining structures (necessary for general site development for erosion control, and so on) may be developed with more attractice materials, including terraced plantings.

REFERENCES

1. *Surveying—Principles and Applications*, 2nd ed., Barry Kavanagh and S. J. Bird, Prentice-Hall, Englewood Cliffs, New Jersey, 1989.

2. *Standard Handbook for Civil Engineers*, 3rd ed., Frederick S. Merritt, McGraw-Hill, New York, 1983.

3. *Time-Saver Standards for Site Planning*, Joseph DeChiara and Lee Koppleman, McGraw-Hill, New York, 1984.

4. *Time-Saver Standards for Landscape Architecture*, Charles Harris and Nicholas Dines, McGraw-Hill, New York, 1988.

5. *Simplified Design of Building Foundations*, 2nd ed., James Ambrose, Wiley, New York, 1988.

6. *Simplified Site Design*, James Ambrose and Peter Brandow, Wiley, New York, 1991.

GLOSSARY

The material presented here consists of a dictionary of the major words and terms from the field of site engineering used in the work in this book. For fuller explanation of most entries, the reader should use the Index to find the related discussions in the text.

Backsight. In levelling, the reading of a height at a point of known elevation. *See* **Foresight.** (Figure 8.1)

Batter boards. Horizontal boards attached to driven stakes to be used as temporary fixed references for site dimensions and/or elevations.

Bearing of a line. Angle of a line in a horizontal plane with reference to north.

Bench mark. Permanent point of known and recorded elevation.

Catch basin. Water-holding device (well, tank, etc.) used to intercept water before it enters a sewer.

Chain. Antique measuring device, consisting of links of heavy wire.

Closed traverse. *See* **Traverse.**

Compass. Device used to establish the direction of north.

Contour. Line on a map connecting points of the same elevation (vertical height).

Contour interval. Difference in height of adjacent contour lines on a map.

Cut. Reduction of the surface of a site below the original grade level. *See* **Fill.**

Datum. The reference level (horizontal plane) to which point elevations or contours are related, usually mean sea level.

Declination, magnetic. At a particular location on the earth's surface, the angle between true north and magnetic (compass) north.

Deflection angle. The angle between a chord and a tangent intersecting the chord at one end; equals one-half the angle subtended by the chord. (Figure 7.5)

Departure. Projection (component) of a line on the east-west line. *See* **Latitude.** (Figure 6.9)

Elevation. Vertical distance from a datum plane (horizontal reference plane). *See* Datum.

Excavation. Cut below the existing grade of a site for purposes of some construction.

Fill. Buildup of the site surface above the original grade. *See* **Cut.**

Footing. A pad, usually of concrete, used to spread a load onto the ground surface, usually placed some distance below the finished grade of the site.

Foresight. In levelling, the reading of a height at a point of unknown elevation. *See* **Backsight.** (Figure 8.1)

Grade. 1. Elevation of the ground surface at some location; often qualified as *original* grade, *finished* (re-established) grade, and so on. 2. Slope of the ground surface (angle from the horizontal) at some location, also called **Gradient** or **Slope.**

Gradient. *See* **Grade.**

Grading. Site work consisting of redevelopment of the form of the ground surface.

Hachures. Short lines drawn perpendicular to a contour line to indicate a depression of the ground surface at the bottom of a slope. (Figure 9.4*b*)

Interpolation. Establishment of a numerical value by linear proportionality between two sequential numbers. Example: finding the elevation of a point by using the adjacent contour elevations and the distances of the point from the contour lines.

Inverse levelling. Determination of the elevation of the underside of something, such as a ceiling or the bottom surface of a bridge. (Figure 8.5)

Invert. Elevation of the lowest part of the inside of a pipe or tunnel.

Latitude. Projection (component) of a line on the north-south line. *See* **Departure.** (Figure 6.9)

Laying off angles. Use of a level to establish the angle of a line with respect to a known line by rotation of the level.

Laying out. Refers to the establishing of the location for some object, such as a foundation or roadway.

Layout. The graphic display (plan, site map, etc.) of the arrangement of some object or objects.

Level. 1. Horizontally flat (zero slope). 2. Elevation (height above a datum plane). 3. Instrument for reading angles in a horizontal plane.

Levelling. The process of determining the difference in elevation between two points, usually by using a **Level** and a **Levelling rod.**

Levelling rod. Rod with accurate graduation markings, used with a **Level** to perform **Levelling.**

Nomograph. An alignment chart used to relate two variables for the determination of a third factor. (Figure 12.2)

Oblique triangle. A triangle without a right (90°) angle.

Obtuse angle. Angle at a corner of a triangle that is larger than 90°.

Open traverse. *See* **Traverse.**

Plane surveying. Surveying that assumes the surface of the earth in a relatively small region to be flat.

Planimeter. Instrument used for measuring the area of plane figures.

Plot. A site defined by established boundaries.

Prismoid. A solid with parallel but unequal ends, whose sides are quadrilaterals or triangles. (Figure 11.3)

Protractor. Instrument for measuring angles on a plane (two-dimensional) figure.

Pythagorean theorem. In any right triangle the square of the hypothenuse is equal to the sum of the squares of the other two sides.

Right triangle. Triangle with one 90° angle.

Runoff. Water flowing from a surface during precipitation (rain).

Simpson's rule. Basis for approximate determination of the area of an irregular plot. (Section 6.7)

Site. A specific point or small region on the surface of the earth.

Slope. The vertical angle between the horizontal and some portion of the ground surface in a particular direction. *See also* **Grade.**

Spirit level. Fluid-filled tube with an air bubble used to establish a horizontal reference.

Spot elevation. Height (elevation; vertical distance) of a point on the ground surface with respect to a **Datum** plane.

Staking out. Process of using stakes driven into the ground surface of a site to locate the position of some construction work.

Survey. 1. General: to observe and note facts about something, such as the general features of a site. 2. To perform surveying work (levelling, etc.). 3. The documentation of surveying work for a site.

Tape. For surveying work: a flexible steel strip with graduated markings for performing linear measurement.

Telescope. Optical device for viewing of distant objects.

Transit. Surveying instrument for measuring both horizontal and vertical angles. *See* **Level.**

Traverse. A line or a series of connected lines surveyed across the earth's surface. A **closed traverse** begins and ends at the same point. An **open traverse** begins at a given point and ends at some distant point.

Vernier. A supplementary scaling device used to determine points intermediate between marked graduations on another scale.

APPENDIX

SOIL PROPERTIES AND BEHAVIORS

The materials in this chapter present a general summary of concerns for the soils that constitute the surface and subsurface ground mass of most sites. This material has been abstracted from various references, but mostly from *Simplified Design of Building Foundations* (Ref. 5).

Information about the materials that constitute the earth's surface comes from a number of sources. Persons and agencies involved in fields such as agriculture, landscaping, highway and airport paving, waterway and dam construction, and the basic earth sciences of geology, mineralogy, and hydrology have generated research and experience that is useful to those involved in general site engineering.

A.1 SOIL CONSIDERATIONS RELATED TO SITE ENGINEERING

Some of the fundamental properties and behaviors of soils related to concerns in site engineering are the following:

Soil Strength

A major concern for bearing-type foundations is the soil resistance to vertical compression. Resistance to horizontal pressure and to sliding friction are also of concern for situations involving the lateral (horizontally directed) effects due to wind, earthquakes, or retained soil.

Dimensional Stability

Soil volumes are subject to change, the principal causes being changes in stress or water content. This affects settlement of foundations and pavements, swelling or shrinking of graded surfaces, and general movements of site structures.

General Relative Stability

Frost actions, seismic shock, organic decomposition, and disturbance during site work can also produce changes in the physical conditions of soils. The degree of sensitivity of soils to these actions is called their relative stability. Highly sensitive soils may require modification as part of the site development work.

Uniformity of Soil Materials

Soil masses typically occur in horizontally stratified layers. Individual layers vary in their composition and thickness. Conditions can also vary considerably at different locations on a site. A major early investigation that must precede any serious engineering design is that for the soil profiles and properties of individual soil components at the site. Depending on the site itself and the nature of the work proposed for the site, this investigation may need to proceed to considerable depth below grade.

Groundwater Conditions

Various conditions, including local climate, proximity to large bodies of water, and the relative porosity (water penetration resistance) of soil layers, affect the presence of water in soils. Water conditions may affect soil stability, but also relate to soil drainage, excavation problems, need for irrigation, and so on.

Sustaining Plant Growth

Where site development involves considerable new planting, the ability of surface soils to sustain plant growth and respond to irrigation systems must be considered. Existing surface soils must often be modified or replaced to provide the necessary conditions.

The discussions that follow present various issues and relationships that affect these and other concerns of site development.

A.2   GROUND MATERIALS

Soil and Rock

The two basic solid materials that constitute the earth's crust are soil and rock. At the extreme, the distinction between the two is clear—loose sand versus solid granite, for example. A precise division is somewhat more difficult, however, since some highly compressed soils may be quite hard, while some types of rock are quite soft or contain many fractures, making them relatively susceptible to disintegration. For practical use in engineering, soil is generally defined as any material consisting of discrete particles that are relatively easy to separate, while rock is any material that requires considerable brute force to excavate.

Fill

In general, fill is material that has been deposited on the site to build up the ground surface. Many naturally occurring soil deposits are of this nature, but for engineering purposes, the term fill is mostly used to describe *man-made fill* or other deposits of fairly recent origin. The issue of concern for fill is primarily its recent origin and the lack of stability that this represents. Continuing consolidation, decomposition, and other changes are likely. The uppermost soil materials on a site are likely to have the character of fill, man-made or otherwise.

Organic Materials

Organic materials near the ground surface occur mostly as partially decayed plant materials. These are highly useful for sustaining new plant growth, but generally represent undesirable stability conditions for various engineering purposes. Organically rich surface soils (generally called *topsoil*) are a valuable resource for landscaping, and may have to be imported to the site where they do not exist in sufficient amounts. For support of pavements, site structures, or building foundations, however, they are generally undesirable.

Investigation of site conditions is done partly to determine the general inventory of these and other existing materials, with a view towards the general management of the site materials for the intended site development. Critical concerns for this management are discussed in Chapter 15.

A.3 SOIL PROPERTIES AND IDENTIFICATION

A typical soil mass is visualized as consisting of three parts, as shown in Figure A.1. The total soil volume is taken up partly by the solid particles and partly by the open spaces between the particles, called the void. The void is typically filled partly by liquid (usually water) and partly by gas (usually air). There are several soil properties that can be expressed in terms of this composition, such as the following.

Soil Weight (γ)

Most of the materials that constitute the solid particles in ordinary soils have a unit density that falls within a narrow range; expressed as specific gravity, the range is from 2.60 to 2.75. Sands typically average about 2.65, clays about 2.70. Notable exceptions are soils containing large amounts of organic mate-

rials. Specific gravity refers to a comparison of the density to that of water, usually considered to weigh 62.4 lb/ft³ [1000 kg/m³]. Thus, for a dry soil sample the soil weight may be determined as follows:

soil unit weight

$$= (\% \text{ of solids})(\text{specific gravity})$$

$$(\text{unit weight of water})$$

Thus for a sandy soil with a void of 30%, the weight may be approximated as follows:

$$\text{soil weight} = \gamma = \left(\frac{70}{100}\right)(2.65)(62.4)$$

$$= 116 \text{ lb/ft}^3$$

$$[1858 \text{ kg/m}^3]$$

Void Ratio (e)

Instead of expressing the void as a percentage, as was done in the preceding example, the term generally used is the void ratio, e, which is defined as follows:

$$e = \frac{\text{volume of the void}}{\text{volume of the solid}}$$

In practice, the void ratio is often determined by using the relationship between soil weight, specific gravity of the sol-

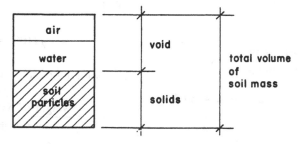

FIGURE A.1.   Three-part composition of a soil mass.

ids, and percentage of the void, as follows. If

$$\gamma = \frac{\text{measured dry weight of the sample}}{\text{measured volume of the sample}}$$

$$= 116 \text{ lb/ft}^3$$

then, assuming a specific gravity (G_s) of 2.65,

$$\gamma = \frac{\% \text{ of solids}}{100} (2.65)(62.4)$$

$$= 116 \text{ lb/ft}^3$$

$$\% \text{ of solids} = \frac{116}{(2.65)(62.4)} (100)$$

$$= 70\%$$

$$\% \text{ of void} = 100 - 70 = 30\%$$

and with the volume expressed as a percentage,

$$e = \frac{\text{volume of the void}}{\text{volume of the solid}} = \frac{30}{70} = 0.43$$

Porosity (n)

The actual percentage of the void is expressed as the porosity of the soil, which in coarse-grained soils (sands and gravels) is generally an indication of the rate at which water flows through or drains from the soil. The actual water flow is determined by standard tests, however, and is described as the relative *permeability* of the soil. Porosity, when used, is simply determined as

n(in percent)

$$= \frac{\text{volume of the void}}{\text{total soil volume}} (100)$$

Thus for the preceding example, $n = 30\%$.

Water Content (w)

The amount of water in a soil sample can be expressed in two ways: by the water content (w) and by the saturation (S). They are defined as follows:

w (in percent)

$$= \frac{\text{weight of water in the sample}}{\text{weight of solids in the sample}}$$

$$\times (100)$$

The weight of the water is simply determined by weighing the wet sample and then drying it to find the dry weight. The saturation is expressed in a ratio, similar to the void ratio, as follows:

$$S = \frac{\text{volume of water}}{\text{volume of void}}$$

Full saturation $(S = 1.0)$ thus occurs when the void is totally filled with water. Oversaturation $(S > 1.0)$ is possible in some soils when the water literally floats some of the solid particles, increasing the void above that in the partly saturated soil mass. In the preceding example, if the soil weight of the sample as taken at the site was found to be 125 lb/ft^3, the water content and saturation would be as follows:

weight of water

$$= (\text{original sample weight})$$

$$- (\text{dry weight})$$

$$= 125 - 116 = 9 \text{ lb/ft}^3$$

Then,

$$w = \tfrac{9}{116} (100) = 7.76\%$$

The volume of water may be found as

$$V_w = \frac{\text{weight of water in sample}}{\text{unit weight of water}}$$

$$= \frac{9}{62.4} = 0.144 \text{ or } 14.4\%$$

Then,

$$S = \frac{14.4}{30} = 0.48$$

The size of the discrete particles that constitute the solids in a soil is significant with regard to the identification of the soil and the evaluation of many of its physical characteristics. Most soils have a range of particles of various sizes, so the full identification typically consists of determining the percentage of particles of particular size categories.

The two common means for measuring grain size are by sieve and sedimentation. The sieve method consists of passing the pulverized dry soil sample through a series of sieves with increasingly smaller openings. The percentage of the total original sample retained on each sieve is recorded. The finest sieve is a No. 200, with openings of approximately

0.003 in. [0.075 mm]. A broad distinction is made between the total amount of solid particles that pass the No. 200 sieve and those retained on all the sieves. Those passing are called the *fines* and the total retained is called the *coarse fraction*.

The fine-grained soil particles are subjected to a sedimentation test. This consists of placing the dry soil in a sealed container with water, shaking the container, and measuring the rate of settlement of the particles. The coarser particles will settle in a few minutes; the finest will take several days.

Figure A.2 shows a graph that is commonly used to record the grain size characteristics for soils. A log scale is used for the grain size, since the range is quite large. The common soil names, based on grain size, are given at the top of the graph. These are approximations, since some overlap occurs at the boundaries, particularly for the fines. The dis-

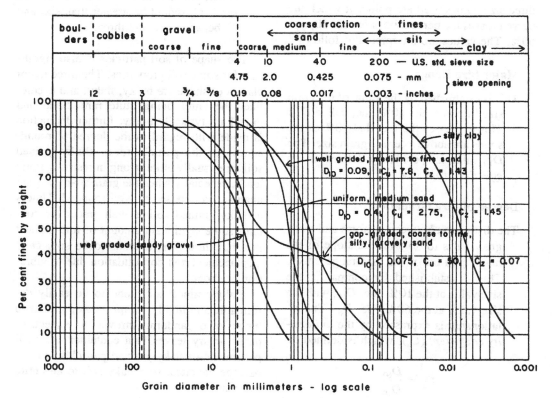

FIGURE A.2.   Grain size range for soils.

tinction between sand and gravel is specifically established by the No. 4 sieve, although the actual materials that constitute the coarse fraction are sometimes the same across the grain size range. The curves shown on the graph are representative of some particularly characteristic soils, described as follows:

A *well-graded* soil consists of some significant percentages of a wide range of soil particles.

A so-called *uniform* soil has a major portion of the particles grouped in a small size range.

A *gap-graded* soil has a wide range of sizes, but with some concentrations of single sizes and small percentages over some ranges.

These size-range characteristics are specifically established by using some actual numeric values from the size-range graph. The three size values used are points at which the curve crosses the percent lines for 10, 30, and 60%. The values are interpreted as follows.

Major Size Range

This is established by the value of the grain size in mm at the 10% line, called D_{10}. This expresses the fact that 90% of the solids are above a certain grain size. The D_{10} value is specifically defined as the *effective grain size*.

Degree of Size Gradation

The distinction between uniform and well-graded has to do with the slope of the major portion of the size-range curve. This is established by comparing the size value at the 10% line, D_{10}, with the size value at the 60% line, D_{60}. The relationship is expressed by the *uniformity coefficient* (C_u), which is defined as

$$C_u = \frac{D_{60}}{D_{10}}$$

The higher this number, the greater the degree of size gradation.

Continuity of Gradation

The value of C_u does not express the character of the graph between D_{60} and D_{10}; that is, it does not establish whether the soil is gap-graded or well-graded, only that it is graded. To establish this, another property is defined, called the *coefficient of curvature* (C_z), that uses all three size values, D_{60}, D_{30}, and D_{10}, as follows:

$$C_z = \frac{(D_{30})^2}{(D_{10})(D_{60})}$$

These coefficients are used only for classification of the coarse-grained soils: sand and gravel. For a well-graded gravel, C_u should be greater than four, and C_z between one and three. For a well-graded sand, C_u should be greater than six, and C_z between one and three.

The shape of soil particles is also significant for some soil properties. The three major classes of shape are bulky, flaky, and needle-like, the latter being quite rare. Sand and gravel are typically bulky; further distinction is made with regard to the degree of roundedness of the particle form. Bulky-grained soils are usually quite strong in resisting static loads, especially when the grain shape is quite angular, as opposed to well rounded. Unless a bulky-grained soil is well-graded or contains some significant amount of fine-grained material, however, it tends to be subject to displacement and consolidation due to vibration or shock.

Flaky-grained soils tend to be easily deformable and highly compressible, similar to the action of randomly thrown loose sheets of paper or dry leaves in a container. A small percentage of flaky-grained particles can impart the character of a flaky soil to an entire soil mass.

Water has various effects on soils, depending on the proportion of water and on the particle shape, size, and chemical properties. A small amount of water tends to make sand particles stick together somewhat. As a result, the sand behaves differently than usual, no longer acting as a loose, flowing mass. When saturated, however, most sands behave like viscous fluids, moving easily under stress due to gravity or other sources. The effect of the variation of water content is generally more dramatic in fine-grained soils. These will change from rocklike solids when totally dry to virtual fluids when supersaturated.

Table A.1 describes the Atterberg limits for fine-grained soils, which are the water content limits, or boundaries, between four stages of structural character of the soil. An important property of such soils is the *plasticity index* (I_p), which is the numeric difference between the liquid limit and plastic limit. A major physical distinction between clays and silts is the range of the plastic state, referred to as the relative plasticity of the soil. Clays have a considerable plastic range and silts

generally have practically none, going almost directly from the semisolid to the liquid state. The plasticity chart, shown in Figure A.3, is used to classify clays and silts on the basis of two properties, liquid limit and plasticity. The line on the chart is the classification boundary between the two soil types.

Another water-related property is the relative ease with which water flows through, or can be extracted from, the soil mass. Coarse-grained soils tend to be rapid draining, or permeable. Fine-grained soils tend to be nondraining, or impervious, and may literally seal out flowing water.

Soil structure may be classified in many ways. A major distinction is that made between soils that are considered to be *cohesive* and those considered *cohesionless*. Cohesionless soils are those consisting predominantly of sand and gravel with no significant bonding of the discrete soil particles. The addition of a small amount of fine-grained material will cause the cohesionless soil to form a weakly bonded mass when dry, but the bonding will virtually disappear with a small percentage of

TABLE A.1 Atterberg Limits for Water Content in Fine-Grained Soils

Description of Structural Character of the Soil Mass	Analogous Material and Behavior	Water Content Limit
Liquid	Thick soup; flows or is very easily deformed	
		Liquid limit: w_L
Plastic	Thick frosting or toothpaste; retains shape, but is easily deformed without cracking	Magnitude of range is *plasticity index*: I_p
		Plastic limit: w_P
Semisolid	Cheddar cheese or hard caramel candy; takes permanent deformation but cracks	
		Shrinkage limit: w_S (Least volume attained upon drying out)
Solid	Hard cookie; crumbles up if deformed	

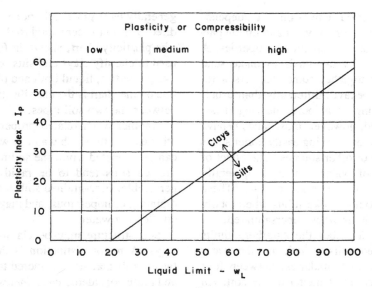

FIGURE A.3.   Plasticity chart using Atterberg limits.

moisture. As the percentage of fine materials is increased, the soil mass becomes progressively more cohesive, tending to retain some defined shape right up to the fully saturated, liquid consistency.

The extreme cases of cohesive and cohesionless soils are typically personified by a pure clay and a pure (or clean) sand, respectively. Typical soil mixtures will range between these two extremes, so they are useful in establishing the boundaries for classification. For a clean sand the structural nature of the soil mass will be largely determined by three properties: the particle shape (well-rounded versus angular), the nature of size gradation (well-graded, gap-graded, or uniform), and the density or degree of compaction of the soil mass.

The density of a sand deposit is related to how closely the particles are fit together and is essentially measured by the void ratio. The actions of water, vibration and shock, and compressive force will tend to pack the particles into tighter arrangements. Thus, the same sand particles may produce strikingly different soil deposits as a result of density variation.

Table A.2 gives the range of density classifications that are commonly used in describing sand deposits, varying from very loose to very dense. The general character of the deposit and the typical range of usable bearing strength are shown as they relate to the density. As mentioned previously, however, the effective nature of the soil depends on additional considerations, principally the particle shape and the size gradation.

Also of concern are the absolute particle size, generally established by the D_{10} property, and the amount of water present, as measured by w or S. A minor amount of water will often tend to give a slight cohesiveness to the sand, as the surface tension in the water partially bonds the discrete sand particles. When fully saturated, however, the sand particles are subject to a buoyancy effect that can work to substantially reduce the stability of the soil.

Many physical and chemical properties affect the structural character of clays. Major considerations are the particle size, the particle shape, and whether the particles are inorganic or organic. The percentage of water in a clay has a very significant effect on its structural nature, changing it from a rocklike material when dry to a viscous fluid when satu-

TABLE A.2 Average Properties of Cohesionless Soils

Relative Density	Blow Count, N (blows/ft)	Void Ratio, e	Simple Field Test with $\frac{1}{2}$-in. Diameter Rod	Usable Bearing Strength (k/ft^2)	(kPa)
Loose	< 10	0.65–0.85	Easily pushed in by hand	0–1.0	0–50
Medium	10–30	0.35–0.65	Easily driven in by hammer	1.0–2.0	50–100
Dense	30–50	0.25–0.50	Driven in by repeated hammer blows	1.5–3.0	75–150
Very dense	> 50	0.20–0.35	Barely penetrated by repeated hammer blows	2.5–4.0	125–200

rated. The property of a clay corresponding to the density of sand is its consistency, varying from very soft to very hard. The general nature of clays and their typical usable bearing strengths as they relate to consistency are shown in Table A.3.

Another major structural property of fine-grained soils is relative plasticity. This was discussed in terms of the Atterberg limits and the classification was made using the plasticity chart shown in Figure A.3. Most fine-grained soils contain both silt and clay, and the predominant character of the soil is evaluated in terms of various measured properties, most significant of which is the plasticity index. Thus, identification as "silty" usually indicates a lack of plasticity (crumbly, fria-

ble, etc.), while that of "claylike" or "clayey" usually indicates some significant degree of plasticity (moldable even when only partly wet).

Various special soil structures are formed by actions that help produce the original soil deposit or work on the deposit after it is in place. Coarse-grained soils with a small percentage of fine-grained material may develop arched arrangements of the cemented coarse particles resulting in a soil structure that is called *honeycombed*. Organic decomposition, electrolytic action, or other factors can cause soils consisting of mixtures of bulky and flaky particles to form highly voided soils that are called *flocculent*. The nature of formation of these soils is shown in Figure A.4. Water de-

TABLE A.3 Average Properties of Cohesive Soils

Consistency	Unconfined Compressive Strength (k/ft^2)	Simple Field Test by Handling of an Undisturbed Sample	Usable Bearing Strength k/ft^2	kPa
Very soft	< 0.5	Oozes between fingers when squeezed	0	0
Soft	0.5–1.0	Easily molded by fingers	0.5–1.0	25–50
Medium	1.0–2.0	Molded by moderately hard squeezing	1.0–1.5	50–75
Stiff	2.0–3.0	Barely molded by strong squeezing	1.0–2.0	50–100
Very stiff	3.0–4.0	Barely dented by very hard squeezing	1.5–3.0	75–150
Hard	4.0 or more	Dented only with a sharp instrument	3.0+	150+

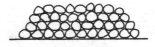

dense, well-compacted soil

loose, compactible soil

honeycombed soil

(a) Cohesionless Soils

oriented, well dispersed
soil formation

partly flocculent
soil formation

highly flocculent
soil formation

(b) Mixed-grain Soils

FIGURE A.4. Arrangements of particles in various soil structures.

posited silts and sands, such as those found at the bottom of dry streams or ponds, should be suspected of this condition if the tested void ratio is found to be quite high.

Honeycombed and flocculent soils may have considerable static strength and be quite adequate for foundation purposes as long as no unstabilizing effects are anticipated. A sudden, unnatural increase in the water content or significant vibration or shock may disturb the fragile bonding, however, resulting in major consolidation of the soil. This can produce major settlement of ground surfaces or foundations if the affected soil mass is large.

Behavior under stress is usually quite different for the two basic soil types: sand and clay. Sand has little resistance to stress unless it is confined. Consider the difference in behavior of a handful of dry sand and sand rammed into a strong container. Clay, on the other hand, has resistance to tension in its natural state all the way up to its liquid consistency. If a hard, dry clay is pulverized, however, it becomes similar to loose sand until some water is added.

In summary, the basic nature of structural behavior and the significant properties that affect it for the two soil types are as follows.

Sand

Little compression resistance without some confinement; principal stress mechanism is shear resistance (interlocking particles grinding together); important properties are angle of internal friction (ϕ), penetration resistance (N) to a driven object such as a soil sampler, unit density in terms of weight or void ratio, grain shape, predominant grain size and nature of size gradation. Some reduction in capacity with high water content.

Clay

Principal stress resistance in tension; confinement generally of concern only in soft, wet clays (to prevent flowing or oozing of the mass); important properties are the unconfined compressive strength (q_u), liquid limit (w_L), plastic index (I_p), and relative consistency (soft to hard).

We must remind the reader that these represent the cases for pure clay and clean sand, which generally represent the outer limits for the range of soil types. Soil deposits typically contain some percentage of all three basic soil ingredients: sand, silt, and clay. Thus, most soils are neither totally cohesive nor totally cohesionless and possess some of the characteristics of both of the basic extremes.

Soil classification or identification must deal with a number of properties for precise categorization of a particular soil sample. Many systems exist and are used by various groups with different concerns. The three most widely used systems are the triangular textural system used by the U.S. Department of Agriculture; the AASHO system, named for its developer, the American Association of State Highway Officials; and the so-called unified system, which is primarily used in foundation engineering. Each of these systems reflects some of the primary concerns of the developers of the system.

The unified system relates to properties of major concern in stress and deformation behavior, excavation and dewatering problems, stability under load, and other issues of concern of foundation designers. The triangular textural system relates to problems of erosion, water retention, ease of cultivation, and so on. The AASHO system relates primarily to the effectiveness of soils for use as base materials for pavements, both as natural soil deposits and as fill material. While the unified system is of major interest in foundation design, there is some value in being familiar with the other systems. A major reason is that in some situations information about sites may be available from highway or agricultural agencies and may be found useful for preliminary analysis and design or for more intelligently determining the nature and extent of soil exploration required for the site. Thus, the ability to make some translations from one system to the other is useful.

Figure A.5 shows the triangular textural system, which is given in graphic form and permits easy identification of the limits used

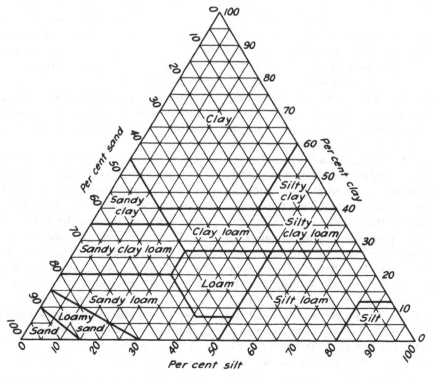

FIGURE A.5. Triangular textural classification chart. From U.S. Department of Agriculture.

to distinguish the named soil types. The property used is strictly grain size percentage, which makes the identification quite approximate since there are potential overlaps between very fine sand and silt and between silt and clay. For agricultural concerns this is of less importance than it may be in foundation engineering. As important as grain size is in broadly characterizing the soil type, there are many more properties that the foundation designer must know, such as grain shape, size gradation, water content, plasticity, and so on.

Use of the textural graph consists of finding the percentages of the three basic soil types and projecting these three points on the edge of the graph to an intersection point that falls in one of the named groups. If the point is near the edge of the group, the soil will have some shared characteristics with the adjacent group. For example, a soil determined to have 46% sand (possibly including some gravel), 21% silt, and 33% clay would fall in the group called "sandy clay loam." However, it would be close to the border of "clay loam" and would be somewhere between the two soil types in actual nature. If a foundation designer were to be told that the soil was somewhere between these two soil types, he could place the location approximately on the triangular graph and project outward to the edges to approximate the percentage of the sand, silt, and clay. That information could be extrapolated to predict a bracketed range of expectations for the behavior of the soil as a foundation material and also to anticipate what additional information would be particularly desirable from a soils investigation.

One useful purpose of the triangular graph is to observe the extent to which percentages of the various materials affect the essential nature of the soil. A sand, for example, must be relatively clean (free of fines) to be considered as such. With as much as 60% sand and only 40% clay the soil is considered essentially a clay. The general situation is that the finer the particles, the greater their influence

on establishing the basic nature of the soil mass on a textural basis.

The AASHO system is shown in Table A.4. The three basic items of data used are the grain size analysis, the liquid limit, and the plasticity index, the latter two properties relating only to fine-grained soils. On the basis of this data the soil type is located by group and its general usefulness as a base for paving is rated, ranging generally from excellent for sand and gravel to poor for clay. This system is also of limited use for foundation design, although it is somewhat more informative than the textural system, especially for fine-grained materials. The AASHO system is also quite useful in evaluating the effectiveness of site materials as bases for floor slabs on grade, sidewalks, and other site paving. As with the textural system, structural properties can only be broadly established and the principal worth is in giving a head start to more extensive soil investigation.

The unified system is shown in Figure A.6. It consists of categorizing the soil into one of 15 groups, each identified by a two-letter symbol. As with the AASHO system, the primary data used are the grain size analysis, the liquid limit, and the plasticity index. It is thus not significantly superior to that system in terms of its data base, but it provides more distinct identification of the soil pertaining to significant considerations of structural behavior.

None of these systems provides information sufficient for foundation design, except for very conservative approximations. What is mainly lacking is any direct testing of the structural properties of the soil (such as penetration resistance of sand or unconfined compression strength of clay), especially in its undisturbed condition at the site. The information obtained from the classification system must therefore be added to that obtained from site observations, site tests, and lab tests for a complete set of data useful for good engineering design.

Building codes and engineering handbooks

TABLE A.4 American Association of State Highway Officials Classification of Soils and Soil Aggregate Mixtures-AASHO Designation M-145

General Classification[a]	Granular Materials (35% or Less Passing No. 200)							Silt-Clay Materials (More than 35% Passing No. 200)			
	A − 1		A − 3	A − 2				A − 4	A − 5	A − 6	A − 7
Group Classification	A − 1 − a	A − 1 − b		A − 2 − 4	A − 2 − 5	A − 2 − 6	A − 2 − 7				A − 7 − 5 / A − 7 − 6

Sieve analysis per cent passing:

	A − 1 − a	A − 1 − b	A − 3	A − 2 − 4	A − 2 − 5	A − 2 − 6	A − 2 − 7	A − 4	A − 5	A − 6	A − 7
No. 10	50 max										
No. 40	30 max	50 max	51 min								
No. 200	15 max	25 max	10 max	35 max	35 max	35 max	35 max	36 min	36 min	36 min	36 min

Characteristics of fraction passing No. 40:

	A − 1 − a	A − 1 − b	A − 3	A − 2 − 4	A − 2 − 5	A − 2 − 6	A − 2 − 7	A − 4	A − 5	A − 6	A − 7
Liquid limit				40 max	41 min	40 max	41 min	40 max	41 min	40 max	41 min
Plasticity index			N.P.[b]	10 max	10 max	11 min	11 min	10 max	10 max	11 min	11 min
Usual types of significant constituent materials	Stone fragments—gravel and sand		Fine sand	Silty or clayey gravel and sand				Silty soils		Clayey soils	
General rating as subgrade	Excellent to good							Fair to poor			

[a]Classification procedure: With required test data in mind, proceed from left to right in chart; correct group will be found by process of elimination. The first group from the left consistent with the test data is correct classification. The A − 7 group is subdivided into A − 7 − 5 or A − 7 − 6 depending on the plastic limit. For w_P < 30, the classification is A − 7 − 6, for $w_P \geqq$ 30, A − 7 − 5.

[b]N.P. denotes nonplastic.

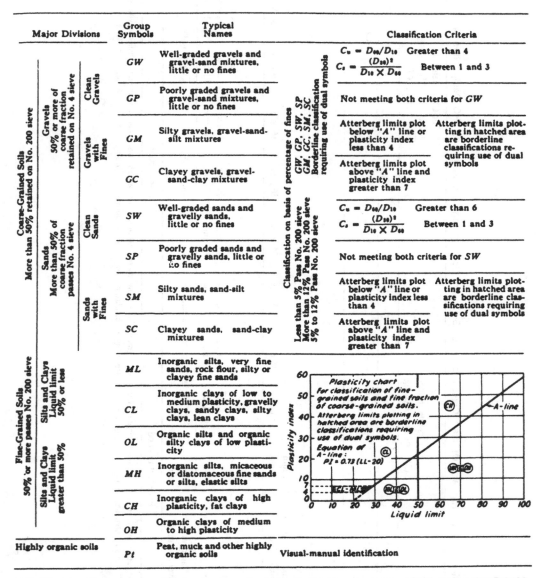

FIGURE A.6. Unified system for classification of soils for engineering purposes, ASTM designation D 2487.

often use some simplified system of grouping soil types for the purpose of regulating or recommending foundation design criteria and some construction details.

A.4 SPECIAL SOIL PROBLEMS

The brief discussion of soil properties and stress behaviors that is given in this book is generally sufficient for many relatively ordinary soil conditions. A great number of special soil situations can be of major concern in foundation design. Some of these are predictable on the basis of regional climate and geological situations. Anyone expecting to do a major amount of foundation design work should study soil mechanics in much greater depth than it is presented in this work. A few of the special problems of particular concern are discussed in the following material.

Expansive Soils

In climates with long dry periods, fine-grained soils often shrink to a minimum volume, sometimes producing vertical cracking in the soil masses that extends to considerable depths. When significant rainfall occurs, two phenomena occur that can produce problems for structures. The first is the swelling of the ground mass as water is absorbed, which can produce considerable upward or sideways pressures on structures. The second is the rapid seepage of water into lower soil strata through the vertical cracks.

The soil swelling can produce major stresses in foundations, especially when it occurs nonuniformly, which is the general case because of paving, landscaping, and so on. Compensation for these stresses depends on the details of the building construction, the type of foundation system, and the relative degree of expansive character in the soil. Local building codes usually have provisions for design with these soils in regions where they are common. If this property is to be expected, it is highly advised that the tests necessary to establish the expansive property be performed, and the results, together with necessary considerations for foundation design, be reviewed by an experienced foundation engineer.

Collapsing Soils

In general, collapsing soils are soils with large voids. The collapse mechanism is essentially one of rapid consolidation, as whatever tends to maintain the soil structure in the large void condition is removed or altered. Very loose sands may display such behavior when they experience drastic changes in water content or are subjected to shock or vibration. The more common cases, however, are those involving soil structures in which fine-grained materials achieve a bonding or molding of cellular

voids. These soil structures may be relatively strong when dry but may literally dissolve when the water content is significantly raised. The bonded structures may also be destroyed by shock or simply by excessive compression stress.

This behavior is generally limited to a few types of soil and can usually be anticipated when such soils display large void ratios. Again, the phenomenon of collapse is often a local condition and is given special consideration in local building codes and practices. The two ordinary methods of dealing with collapsing soils are to stabilize the soil by introducing materials to partly fill the void and reduce the potential degree of collapse, or to use vibration, saturation, or other means to cause the collapse to occur prior to construction. When considerable site grading is to be done, it is sometimes possible to temporarily place soil to some depth on the site, providing sufficient compression to cause significant deformation of the potential foundation-bearing materials. The latter method is effective only for soils in which the static pressure thus produced is truly capable of causing significant consolidation.

A.5   SOIL MODIFICATION

With regard to building construction, modifications of existing soil conditions occur in a number of ways. The recontouring of the site, placing of the building on the site, covering of the ground surface with large areas of pavements, installation of plantings, and installation and continuous operation of irrigation systems all represent major changes in the site environment. These changes, done in a few months or years, are equivalent to ones that may take thousands of years to occur by natural causes, and the sudden disruption of the equilibrium of the geological environment is likely to have an impact that results in readjustments that will eventually cause major changes in soil conditions—particularly those near the ground surface. The *major* modification of soil conditions, therefore, is the construction of the building.

To achieve the building construction, or to compensate for disruptions caused by the construction, or, in some cases, to protect against future potential problems, it is sometimes necessary to make deliberate modifications of soil conditions. The most common type of modification is that undertaken to densify some soil deposit by compaction, preconsolidation, cementation, or other means. This type of modification is usually done either to improve bearing capacity and reduce settlements, or to protect against future failures in the form of subsidence, slippage of slopes, or erosion. Another common modification relates to changes in the groundwater and moisture conditions. Covering of the site surface with construction, installation of irrigation and site runoff drainage systems, and any dewatering done to improve conditions during construction cause changes of this type. Deliberate changes may consist of altering the soil to change its degree of permeability or its water-retaining properties. Fine-grained materials may be leached from a mixed grain soil to make it more cohesionless; fine materials may be introduced into a coarse-grained soil to make it less permeable or to cement the particles into a more stable mass.

In many cases site materials may be removed and other materials brought to the site to replace them. This is often necessary in order to have materials appropriate for fill under slabs or around foundation construction. In some cases it may be possible to use site materials by first removing them and then reinstalling them with some alteration: heavy compaction, added materials, and so on.

In general, it is necessary to look at the site ground materials as construction materials. These materials may be useful or not; may be subject to processing for improvement of their character; may be moved to other locations on the site for some appropriate purpose; or may be removed and traded for different materials of a more useful nature. In this regard, the preliminary site soil investigation may be

viewed as an inventory of potential construction materials—a gift from nature. When nature is not so generous, the major techniques for correction, other than the construction of the building and general development of the building site, are the following.

Removal and Replacement

Existing surface materials are quite often removed and replaced, even though the existing surface contours are not to be altered substantially. Soil materials more desirable for plant growth or as base materials for paving may be brought to the site and deposited. Occasionally, it is possible to improve the bearing condition immediately beneath footings by removing a relatively thin layer of soil and replacing it with a layer of highly compacted gravel. For relatively small footings, where the magnitude of vertical pressure is rapidly dissipated below the bottom of the footing, this process may permit bearing on an otherwise marginally acceptable soil deposit. Another situation of this type is that where the soil at the level of the bottom of the foundation construction is difficult to use as a working surface during the construction work. In the latter case, the general level of the bottom of the excavation may be cut somewhat lower and a layer of fine gravel placed to provide a better surface. Dewatering of the construction site may also be possible as an alternative to soil removal in some situations.

Compaction

Various techniques can be used to lower the void in a soil deposit. For surface soils the common methods use some combination of wetting, rolling, tamping (pounding), and use of various machinery and equipment to consolidate the soil to create a denser surface or a better substratum condition for a pavement base. Flooding of an excavation is sometimes done to soften the soil bonds in bearing materials with a high void, resulting in collapse that may otherwise occur when extensive irrigation of the site occurs at a later time. Sometimes holes are drilled into lower soil strata to increase water penetration. Where extensive grading is done or where surface soils are saved for resurfacing after construction is complete, large masses of soil may be stacked on a building site to provide a massive weight to consolidate soils near the ground surface. For very loose cohesionless soils, vibration may be effective in reducing void created by stacked soil particles (see situations illustrated in Figure A.4).

Infiltration

For some types of soils, notably loose sands, an improvement is sometimes achieved by adding a filler material to the soil to fill part of the void and possibly to provide a bonding of the loose particles. Portland cement and bentonite are two types of very fine-grained materials that can be made to enter the void of soils of larger particle size to achieve these effects.

Except for relatively simple compaction of surface materials, all soil modifications should be undertaken only with the advice of an experienced soils consultant.

A.6 DESIGN CRITERIA

For the design of ordinary bearing-type foundations, several structural properties of a soil must be established. The principal values are the following:

Allowable Bearing Pressure

This is the maximum permissible value for vertical compression stress at the con-

tact surface of bearing elements. It is typically quoted in units of pounds or kips per square foot of contact surface.

Compressibility

This is the predicted amount of volumetric consolidation that determines the amount of settlement of the foundation. Quantification is usually done in terms of the actual dimension of vertical settlement predicted for the foundation.

Active Lateral Pressure

This is the horizontal pressure exerted against retaining structures, visualized in its simplest form as an equivalent fluid pressure. Quantification is in terms of a density for the equivalent fluid given in actual unit weight value or as a percentage of the soil unit weight.

Passive Lateral Pressure

This is the horizontal resistance offered by the soil to forces against the soil mass. It is also visualized as varying linearly with depth in the manner of a fluid pressure. Quantification is usually in terms of a specific pressure increase per unit of depth.

Friction Resistance

This is the resistance to sliding along the contact bearing face of a footing. For cohesionless soils, it is usually given as a friction coefficient to be multiplied by the compression force. For clays, it is given as a specific value in pounds per square foot to be multiplied by the contact area.

Calculation of settlements is quite complex and is beyond the scope of this book. It is best to be generally aware of the conditions that may produce settlement problems, but the actual quantification of settlement values should be done by an experienced soils engineer.

Whenever possible, stress limits should be established as the result of a thorough investigation and the recommendations of a qualified soils engineer. Most building codes allow for the use of so-called *presumptive* values for design. These are average values, on the conservative side usually, that may be used for soils identified by groupings used by the codes.

Table A.5 presents a summary of information for the basic soil types classified by the unified system. Information is grouped as follows.

Significant Properties

These are the properties that relate directly to the identity or stress-and-strain evaluation of the soil.

Simple Field Identification

These are the various characteristics of the soil that may be used for identification in the absence of testing. They may also be used to verify that the soil encountered during construction is that indicated by the soil exploration and assumed for the foundation design.

Average Properties

These are approximate values and are best verified by exploration and testing, but should be adequate for preliminary design. Stress values are generally those in the approximate range recommended by building codes.

A.7 FILL

Site development typically involves the use of soils for fill. This may involve situations in which some site areas are to be raised to a new finish surface (see Figure A.7a), for which a general profile is developed by filling depressions with soil moved from other areas on the site that are being cut down to a new

grade. If more filling than cutting is required on the site, or surface soils are unsatisfactory, fill materials will have to be brought to the site.

In most cases, selection of fill soils and their handling must be controlled to some degree. A major concern is to maintain the finished grade profile once construction is completed. This is of much greater concern if pavements or other site structures are to be placed on the fill. In general, it is not feasible—or even allowed by codes—to place building foundations on fill.

To avoid settlement of the surface over fills (called surface *subsidence*), it is necessary to use some methods to compact the fill materials. This is usually done with equipment, working with thin layers of the fill in subsequent built-up accumulation to achieve the necessary finished grade.

The equipment and procedures used for compaction depend on the type of soil, the extent of the area being compacted, and the degree of soil volume reduction desired. The latter is defined in terms of some desired degree of density compared to that of a soil with zero void. Since actual zero void is hardly feasible, a density representing from 90 to 95% of a fully compacted volume is usually acceptable as significant compaction.

The types of soils that yield good compaction and have other properties desirable for fills are limited generally to those with predominantly coarse-grained materials and small amounts of fine-grained materials (silty sands, gravel with traces of silt and clay, etc.). Where fast draining of the fill is a requirement, the amount of fine-grained materials may be strictly limited. When soils at the site do not have these properties, they may be modified; otherwise, imported materials must be used.

Backfill

A special situation is that where soil is used to fill in around construction, where overcutting of the excavation has been required (see

TABLE A.5 Summary of Properties and Recommended Design Values for Soils Classified by the Unified System (ASTM Designation D-2487)

Description	Gravel, well-graded; little or no fines			Gravel, poorly graded, little or no fines			Silty gravel and gravel–sand–silt mixes		
ASTM Classification (See Figure A.6)	GW			GP			GM		
Significant properties	≥95% retained on No. 200 sieve (0.003 in.) ≥50% of coarse fraction retained on No. 4 sieve ($\frac{3}{16}$ in.) $C_u > 4$ $1 < C_z < 3$			≥95% retained on No. 200 sieve (0.003 in.) ≥50% of coarse fraction retained on No. 4 sieve ($\frac{3}{16}$ in.) Does not meet C_u and/or C_z requirements for well-graded (GW)			50–88% retained on No. 200 sieve (0.003 in.) ≥50% of coarse fraction retained on No. 4 sieve ($\frac{3}{16}$ in.) Atterberg plot below A line or $I_p < 4$		
Field Identification	Significant amounts of coarse rock fragments; easily pulverized; fast draining; wide range of grain sizes			Significant amounts of coarse rock fragments; easily pulverized; fast draining; has narrow range of sizes or is gap-graded			Gravely but forms clumps that pulverize with moderate effort; wet sample takes little or no remolding before disintegrating; slow draining		
Average Properties	Loose ($N < 10$)	Medium ($10 < N < 30$)	Dense ($N > 30$)	Loose ($N < 10$)	Medium ($10 < N < 30$)	Dense ($N > 30$)	Loose ($N < 10$)	Medium ($10 < N < 30$)	Dense ($N > 30$)
Allowable bearing (lb/ft²) with minimum of 1 ft surcharge	1300	1500	2000	1300	1500	2000	1000	1500	2000
increase for surcharge (%/ft)	20	20	20	20	20	20	20	20	20
maximum total	8000	8000	8000	8000	8000	8000	8000	8000	8000
Lateral pressure									
active coefficient	0.25	0.25	0.25	0.25	0.25	0.25	0.30	0.30	0.30
passive (lb/ft² per ft depth)	200	300	400	200	300	400	167	250	333
Friction (coefficient or lb/ft²)	0.50	0.60	0.60	0.50	0.60	0.60	0.40	0.50	0.50
Weight (lb/ft³)									
dry	100	110	115	90	100	110	100	115	130
saturated	125	130	135	120	125	130	125	135	145
Compressibility	Medium	Low	Very low	Medium	Low	Very low	Medium	Low	Very low

TABLE A.5 (Continued)

	Clayey gravel and gravel–sand–clay mixes			Sand, well-graded; gravely sand; little or no fines			Sand, poorly graded; gravely sand; little or no fines		
Description									
ASTM Classification (See Figure A.6)	GC			SW			SP		
Significant Properties	50–88% retained on No. 200 sieve (0.003 in.) ≧50% of coarse fraction retained on No. 4 sieve ($\frac{3}{16}$ in.) Atterberg plot above A line or $I_p > 7$			≧95% retained on No. 200 sieve (0.003 in.) >50% of coarse fraction passes No. 14 sieve ($\frac{3}{16}$ in.) $C_u > 6$ $C_z\ 1 - 3$			≧95% retained on No. 200 sieve (0.003 in.) >50% of coarse fraction passes No. 4 sieve ($\frac{3}{16}$ in.) Does not meet C_u and/or C_z requirements for well-graded (SW)		
Field Identification	Gravely but forms hard clumps that require considerable effort to pulverize; wet sample takes some remolding before disintegrating; very slow draining			Relatively clean sand with wide size range; easily pulverized; fast draining			Relatively clean sand with narrow size range or gaps in grading; easily pulverized; fast draining		
Average Properties	Loose ($N < 10$)	Medium ($10 < N < 30$)	Dense ($N < 30$)	Loose ($N < 10$)	Medium ($10 < N < 30$)	Dense ($N > 30$)	Loose ($N < 10$)	Medium ($10 < N < 30$)	Dense ($N > 30$)
Allowable bearing (lb/ft²) with minimum of 1 ft surcharge	1000	1500	2000	1000	1500	2000	1000	1500	2000
increase for surcharge (%/ft)	20	20	20	20	20	20	20	20	20
maximum total	8000	8000	8000	6000	6000	6000	6000	6000	6000
Lateral pressure									
active coefficient	0.30	0.30	0.30	0.25	0.25	0.25	0.25	0.25	0.25
passive (lb/ft² per ft depth)	167	250	333	183	275	367	75	150	200
Friction (coefficient or lb/ft²)	0.40	0.50	0.50	0.35	0.40	0.40	0.35	0.40	0.40
Weight (lb/ft³)									
dry	110	120	125	100	110	115	90	100	110
saturated	125	130	135	125	130	135	120	125	130
Compressibility	Medium	Low	Very low	Medium high	Medium	Low	Medium high	Medium	Low

	Silty sand and sand–silt mixes			Clayey sand and sand-clay mixes			Inorganic silt, very find sand, rock flour, silty or clayey fine sand		
ASTM Classification (See Figure A.6)	SM			SC			ML		
Significant Properties	50–80% retained on No. 200 sieve (0.003 in.) >50% of coarse fraction passes No. 4 sieve ($\frac{3}{16}$ in.) Atterberg plot below A line or I_p < 4			50–80% retained on No. 200 sieve (0.003 in.) >50% passes No. 4 sieve ($\frac{3}{16}$ in.) Atterberg plot above A line or I_p > 7			$\geqq$50% passes No. 200 sieve (0.003 in.) $w_L \leqq 50\%$ Atterberg plot below A line or I_p < 20		
Field Identification	Sandy soil; forms clumps that can be pulverized with moderate effort; wet sample takes little remolding before disintegrating; slow draining			Sandy soil; forms clumps that offer some resistance to being pulverized; wet sample takes some remolding before disintegrating; very slow draining			Fine-grained soils of low plasticity; slow draining; dry clumps easily pulverized; won't from thin thread when molded		
Average Properties	Loose (N < 10)	Medium ($10 < N < 30$)	Dense (N > 30)	Loose or soft (N < 10)	Medium ($10 < N < 30$)	Dense or stiff (N > 30)	Loose or soft (N < 10)	Medium ($10 < N < 30$)	Dense or stiff (N < 30)
Allowable bearing (lb/ft²) with minimum of 1 ft surcharge	500	1000	1500	1000	1500	2000	500	750	1000
increase for surcharge (%/ft)	20	20	20	20	20	20	20	20	20
maximum total	4000	4000	4000	4000	4000	4000	3000	3000	3000
Lateral pressure									
active coefficient	0.30	0.30	0.30	0.30	0.30	0.30	0.35	0.35	0.35
passive (lb/ft² per ft depth)	100	167	233	133	217	300	67	100	133
Friction (coefficient or lb/ft²)	0.35	0.40	0.40	0.35	0.40	0.40	0.35 or 250	0.40 or 375	0.40 or 500
Weight (lb/ft³)									
dry	105	115	120	105	115	120	105	115	120
saturated	125	130	135	125	130	135	125	130	135
Compressibility	Medium	Low	Low	Medium	Low	Low	Medium high	Medium	Low

TABLE A.5 *(Continued)*

Description	Lean clay; inorganic clay of low to medium plasticity; gravely clay; sandy clay; silty clay			Organic silt and organic silty clay of low plasticity			Inorganic silt; micaceous or diatomaceous fine sands or silt; elastic silt		
ASTM Classification (See Figure A.6)	CL			OL			MH		
Significant Properties	≧50% passes No. 200 sieve (0.003 in.) $w_L \leqq 50\%$ Atterberg plot above A line or $I_p > 7$			≧50% passes No. 200 sieve (0.003 in.) $w_L \leqq 50\%$ Atterberg plot below A line or $I_p < 20$			≧50% passes No. 200 sieve (0.003 in.) $w_L > 50\%$ Atterberg plot below A line or $I_p < 20$		
Field Identification	Fine-grained soil of low plasticity; slow draining; dry clumps quite hard, but not very difficult to pulverize			Fine-grained soil of low plasticity; slow draining; dry clumps quite hard, but not very difficult to pulverize; typical slight musty, rotting odor			Fine-grained soils of low plasticity; slow draining; dry clumps quite hard, but not very difficult to pulverize; spongy; compressible wet or dry		
Average Properties	Soft	Medium	Stiff	Loose or soft (N < 10)	Medium (10 < N < 30)	Dense or stiff (N > 30)	Loose or soft (N < 10)	Medium (10 < N < 30)	Dense or stiff (N > 30)
Allowable bearing (lb/ft²) with minimum of 1 ft	1000	1500	2000	500	750	1000	500	750	1000
increase of surcharge (%/ft)	20	20	20	10	10	10	10	10	10
maximum total	3000	3000	3000	2000	2000	2000	1500	1500	1500
Lateral pressure active coefficient	0.40	0.50	0.75	0.75	0.85	0.95	0.50	0.60	0.75
passive (lb/ft² per ft depth)	267	467	667	33	50	67	33	50	67
Friction (coefficient or lb/ft²)	500	750	1000	250	375	500	200	300	400
Weight (lb/ft²) dry	80	95	105	75	90	100	70	85	100
saturated	110	120	130	95	105	115	100	110	120
Compressibility	High	Medium high	Low	High	Medium high	Medium	Very high	High	Medium high

Description	Fat clay; inorganic clay of high plasticity			Organic clay of medium to high plasticity			Peat, muck, topsoil
ASTM Classification (See Figure A.6)	CH			OH			Pt
Significant Properties	≧50% passes No. 200 sieve (0.003 in.) $w_L > 50\%$ Atterberg plot above A line or $I_p > 20$			≧50% passes No. 200 sieve (0.003 in.) $w_L > 50\%$ Atterberg plot below A line			Highly organic Low density
Field Identification	Fine-grained soil of high plasticity; sticky and highly moldable without fracture when wet; nondraining; impervious; dry clumps very hard and very difficult to pulverize; highly compressible			Fine-grained soil of medium to high plasticity; stickly and moderately moldable without fracture when wet; nondraining; impervious; dry clumps hard; moderately difficult to pulverize; highly compressible; typical slight musty, rotting odor			Contains large amounts of partially decomposed plant or animal materials; strong rotting odor; slow draining highly compressible
Average Properties	Soft	Medium	Stiff	Soft	Medium	Stiff	
Allowable bearing (lb/ft²) with minimum of 1 ft surcharge	500	750	1000	500	500	500	Not usable
increase for surcharge (%/ft)	10	10	10	0	0	0	
maximum total	1500	1500	1500	500	500	500	
Lateral pressure active coefficient	0.75	0.85	0.95	0.75	0.85	0.95	0.30
passive (lb/ft² per ft depth)	33	100	167	33	33	33	Not usable
Friction (coefficient or lb/ft²)	200	300	400	150	200	200	Not usable
Weight (lb/ft³) dry	75	90	105	65	85	100	70–90
saturated	95	110	130	90	110	125	90–110
Compressibility	Very high	High	Medium high	Very high	High	Medium high	Very high

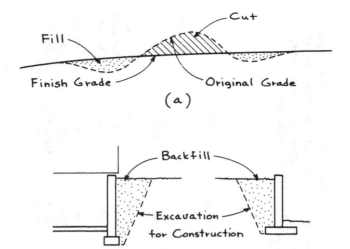

FIGURE A.7.   Use of fill in site construction: (a) for levelling of a hilly site; (b) as backfill for site construction.

Figure A.7*b*). This fill is referred to as *backfill*, and its placement is often controlled for various purposes, including the following:

Prevention of subsidence of the finished grade resulting in undesirable depressions next to the construction.

Prevention of damage to the construction, especially to waterproofing on surfaces of walls.

Development of perimeter drain lines.

Fulfillment of these or other special purposes may affect the choice of fill materials, specifications for compaction, or other details of the installation. The usual, practical thing to do is to use the excavated materials where they can be practically stockpiled nearby for this purpose. However, these soils may not have the desired properties for the backfill purposes, and imported materials may be required.

Fill for Pavements

Development of pavements often involves the use of fill materials. Figure A.8 shows some

of the typical components involved when a pavement must be built up to some level and a stable paved surface is desired. This stacking of conditioned layers usually begins with some treatment of the existing soil surface that has been exposed by cutting of the original grade. This surface layer of soil may be modified by compaction, wetting, filling with some special material to increase density, and so on.

The paving itself may be achieved with a variety of materials, but if it is a hard surfacing of concrete, asphalt, or tiles, its direct support is usually achieved with a coarse-grained soil layer that is compacted. The degree of compaction usually relates to the form of paving and the kind of traffic being borne.

If the paved surface is developed a short distance above the undisturbed soil surface, the directly supporting fill material may be the only base fill required. However, if the surface must be some distance above the existing cut surface, it is probably too expensive to use this special fill, and some intermediate layer of fill may be used. This level-raising fill must also be controlled; again, the degree depends on the type of paving and the traffic.

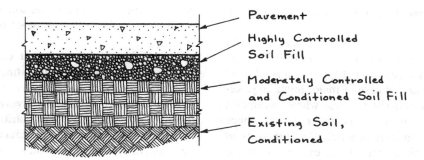

- Pavement
- Highly Controlled Soil Fill
- Moderately Controlled and Conditioned Soil Fill
- Existing Soil, Conditioned

FIGURE A.8. Typical construction for a pavement where the pavement surface is some distance above the existing soil surface.

A further complication for fills under pavements is the usual need for use of soil materials with good draining properties to prevent water buildup under the paving. Where this is especially critical, a piped drainage system may be installed in the supporting fill materials.

A.8 SOILS FOR LANDSCAPING

Full site development often involves the controlled development of all site surfaces. Surfaces not covered by construction are typically covered by either paving or plantings. Development of hard surface paving is discussed in the preceding section; it typically requires the use of some soil materials to develop a built-up base for the actual surface pavement.

Loose Paving Materials

A special form of paving is that achieved with loose soil materials—the dirt path or gravel road, for two examples. These may be quite crudely developed, or more carefully built up as for hard paving, using all the layers shown in Figure A.8 with the controlled loose surfacing replacing the hard pavement. Good surface paving, as well as good sub-base materials, may be found on the site, but this is most likely a fortunate accident, and some

imported materials with the desired properties are typically required. Although typically described as loose, the upper layers of this form of paving are likely to be more durable if some degree of compaction is achieved with them.

Materials used for loose paving depend on the form of traffic and the appearance desired. Foot traffic may be borne on pulverized materials such as bark, brick, or clinkers. A good "dirt" path may be developed with sand and a small amount of silt to promote compaction by the traffic. Vehicular traffic usually requires the use of relatively coarse gravel; again, one containing a controlled amount of fine materials to promote compaction by the traffic itself.

Topsoil

Topsoils exist naturally on many sites where plant growth has occurred for many years. Principal ingredients of what is referred to as topsoil are organic materials, typically consisting largely of partly decomposed plant materials. Although the exact composition of topsoils varies considerably, they generally represent soils that sustain plant growth.

Development of planted areas on sites requires the use of some form of soil that can sustain plant growth. An exact specification for such a soil will relate to the forms of plantings being installed, with special needs for particular nutrients and response to water. Very ordinary topsoils may be adequate in some cases, very highly controlled ones in others.

Where feasible, existing topsoil on a site is typically preserved—left in place if no regrading is required or stripped and stockpiled for relocation in case of major regrading. Removal may also be required where pavements or other constructions are required, since structural support is unreliable with soils of high organic content. If the existing topsoil is not all required on a site, it can sometimes be exported and sold for use in other site situations.

Topsoil for site development can sometimes be produced by special treatment of existing site materials through the use of various additives. Typical additions involve fertilizers or chemicals to improve nutritive properties and various materials to reduce density and improve drainage or moisture retention.

INDEX